THE ARCTIC

CHOICES FOR PEACE AND SECURITY

PROCEEDINGS OF
A PUBLIC INQUIRY

Thomas R. Berger,
Soviet Ambassador Alexei Rodionov,
Douglas Roche
and 21 other speakers

Presented by
The True North Strong & Free
Inquiry Society

Gordon Soules Book Publishers Ltd.
West Vancouver, Canada
Seattle, U.S.A.

Copyright © 1989 by The True North Strong and Free Inquiry Society, Edmonton, Alberta, Canada

All rights reserved. No part of this book may be reproduced in any form by any means without the written permission of the publisher, except by a reviewer, who may quote passages in a review.

Canadian Cataloguing in Publication Data

Main entry under title:

The Arctic : Choices for Peace and Security

Conference held March 18—19, 1989, in Edmonton.
ISBN 0-919574-82-3

1. Arctic regions - National security - Congresses.
2. Peace - Congresses. 3. Environmental policy - Arctic regions - Congresses. 4. Arctic regions - Military policy - Congresses.
I. Berger, Thomas R. II. True North Strong and Free Inquiry Society.
UA600.T72 1989   355'.0330719   C89-091589-X

Published in Canada by
Gordon Soules Book Publishers Ltd.
1352-B Marine Drive
West Vancouver, B.C.
Canada V7T 1B5

Published in the U.S.A. by
Gordon Soules Book Publishers Ltd.
620 - 1916 Pike Place
Seattle, WA 98101

Information regarding programs and activities of The True North Strong and Free Inquiry Society should be addressed to:
The True North Strong and Free Inquiry Society
51110 Range Road 223
Sherwood Park, Alberta, Canada
T8C 1G9

Editorial assistance by Yvonne Van Ruskenveld, Anne Norman and Hilda McKenzie
Typesetting by Joy Woodsworth and The Typeworks
Designed by Chris Bergthorson
Map 1 drawn by Andrew Macpherson and John Winfield
Maps 2, 3 and 4 reproduced from sheet MCR 22, by permission of the Department of Energy, Mines and Resources, Ottawa. Added information by Andrew Macpherson
Printed and bound in Canada by Hignell Printing Limited

Table of Contents

Acknowledgments..5
Introduction by W. H. Hurlburt................................9

1. Opening Remarks 13
   Mary Collins
2. The Arctic: An Overview....................................19
   John F. Merritt
3. Security, Peace and the Native Peoples of the Arctic..........31
   Mary Simon
4. The North as Frontier and Homeland 37
   Thomas R. Berger
5. Responsible Development and Associated Economic,
   Environmental and Social Considerations 45
   Edward R. Weick
6. Multilateral Cooperation to Protect the Arctic Environment: The
   Finnish Initiative 53
   Ambassador Esko Rajakoski
7. Question Session..61
   *Thomas R. Berger, John F. Merritt, Ambassador Esko Rajakoski,
   Mary Simon and Edward R. Weick*
8. The Human Foundation for Peace and Security in the Arctic ...87
   Gordon Robertson
9. International Arctic Cooperation: A Canadian Dream or a
   Necessity?..93
   Walter Slipchenko
10. The Role of International Law for Peace and Security in the
    Arctic 105
    Donat Pharand
11. The Militarization and Security Concept of the Arctic........113
    Willy Østreng
12. Arctic "Militarization": The Canadian Dilemma 127
    Keith R. Greenaway
13. Question Session 139
    *Keith R. Greenaway, Willy Østreng, Donat Pharand, Gordon
    Robertson and Walter Slipchenko*
14. The Arctic, Northern Waters and Arms Control 157
    Johan Jørgen Holst
15. Canadian Circumpolar Relations 169
    Ambassador Douglas Roche

16. Canadian Defence Policies and Activities in the Arctic 175
    Major-General David Huddleston
17. International Relations: A Government of the Northwest
    Territories View. 185
    Dennis Patterson
18. Question Session 193
    *Johan Jørgen Holst, Major-General David Huddleston,
    Dennis Patterson and Ambassador Douglas Roche*
19. Soviet Approaches to Security and Peaceful Cooperation in the
    Arctic: An Overview 209
    Ambassador Alexei A. Rodionov
20. Soviet Approaches to Security in the Arctic 215
    Andrey E. Granovsky
21. Soviet Approaches to Peaceful Cooperation in the Arctic 223
    Yuri V. Kazmin
22. Maritime Strategy and Nuclear-free Zones 227
    Captain James T. Bush
23. Question Session 237
    *Captain James T. Bush, Andrey E. Granovsky, Yuri V. Kazmin
    and Ambassador Alexei A. Rodionov*
24. Canadian Policy Options for the Arctic 245
    Harriet Critchley
25. Take It from the Top: A Proposal for a Nuclear-weapons-free
    Arctic 251
    Steve Shallhorn
26. Question Session 263
    Harriet Critchley and Steve Shallhorn
27. Summation 267
    Gurston Dacks and Franklyn Griffiths
28. Closing Remarks 279
    Brian Sproule

Panelists Gwynne Dyer, Linda Hughes and Ann Medina participated in question sessions in the following chapters: 7, 13, 18 and 23.

Stephen Lewis was the moderator on the first day of the inquiry (chapters 7 and 13).

Adrienne Clarkson was the moderator on the second day of the inquiry (chapters 13, 18, 23 and 26).

Acknowledgments

The Arctic: Choices for Peace and Security—A Public Inquiry, held March 18 and 19, 1989, in Edmonton, Alberta, drew over fifteen hundred participants. They heard speakers from five circumpolar countries and numerous distinguished northerners and noted Canadian commentators discuss the Arctic as a major focal point for international peace and security. The following pages attest to the substance and significance of what was presented. What black and white cannot reproduce is the actual experience. Something magical happened when thoughtful people from virtually every corner of the Northern Hemisphere shared ideas with one another.

The inquiry was organized by The True North Strong and Free Inquiry Society, which formed in 1986 under the leadership of Mel Hurtig, founding member of the Council of Canadians, and Brian Sproule, president of the Edmonton Chapter of Physicians for Social Responsibility (now Canadian Physicians for the Prevention of Nuclear War). Following the success of its November 1986 inquiry into Canadian defence policy and nuclear arms, the society evolved into a charitable society with a mandate to promote the education of the public on matters of general concern, hence "The Arctic: Choices for Peace and Security" was chosen as the theme for the 1989 inquiry. This unique inquiry on the Arctic was successful because of the collective efforts of many individuals and organizations, who contributed their ideas, talents and money.

The inquiry was coordinated by the Board of Directors of The True North Strong and Free Inquiry Society. One member, William Hurlburt, deserves special mention, firstly for the nerve to mastermind such an exceptional line-up of speakers and moderators, and secondly for the

Acknowledgments

audacity to pull them together at the same time in the same place. We are indebted to those speakers and moderators who waived their customary fees and adjusted their schedules so that they could attend the inquiry. The rest of the fearless team who successfully presented this inquiry were as follows: Bill Berezowski, Karen Farkas, Kevin Hanson, Clyde Hurtig, Helen Ready, John Ritchie, Bill Stollery and Jill Wright. The program assistant was Dr. Andrew Macpherson, whose expertise and extensive experience in the North were invaluable. The executive assistant to the board was Connie Morley, and the inquiry assistants were Chris Cavanaugh and Colin Park.

Money makes the world go round and may help to make the true North strong and free. Sponsorship by the following organizations made this inquiry possible: Canadian Institute for International Peace and Security; Department of External Affairs, Government of Canada; The Walter and Duncan Gordon Charitable Foundation; and Canadian Physicians for the Prevention of Nuclear War.

We also acknowledge the generous contributions in kind by the following: Edmonton Convention Centre; Cytronics Technologies, Inc.; Hurtig Publishers Ltd.; Studio 3 Graphics; Co-op Printing; Pièce de Résistance Ltée.; Word and Image; Alberta Display; Spectrum; and John Sproule.

Special thanks must be given to the Bearclaw Gallery, Edmonton, Alberta, which gathered indigenous art from throughout northwestern North America to display at the inquiry, the University of Alberta map collection and the Ministry of Foreign Affairs of the Soviet Union for the map of the Northern Hemisphere and photographic displays. Acknowledgments are also due to Energy, Mines and Resources Canada, on whose maps some of the maps in this book are based.

Thanks also to Mini Aodla Freeman, author of *Life Among the Qallunaat,* who wrote a poem specifically for the ecumenical service at the inquiry, and the Reverend John Marsh who coordinated and led this moving ecumenical service.

Music was composed and produced specifically for the inquiry by Edmonton composer Roger Deegan. Its theme was inspired by an Arctic song originating from the peoples of the Alaskan and Yukon coast.

Shaw Cable enabled the inquiry to be videotaped and broadcast repeatedly. The highlights of the inquiry will be available for purchase in a one-hour video from The True North Strong and Free Inquiry Society.

Thank you also to the sixty-three individuals who contributed financially, the one hundred dedicated volunteers and the thirty-five organizations that endorsed the inquiry. Most of all thanks to the over fifteen hundred people, some of whom came from points far north, to share, listen to and discuss points of view on the Arctic with respect to the role it plays in international peace and security, its peoples and environment, and the

Acknowledgments

complex of relations among circumpolar countries.

We thank those who made the proceedings of the inquiry available in book form, particularly William Hurlburt, Karen Farkas, Andrew Macpherson, who is also responsible for most of the footnotes and maps contained therein, and Jill Wright. Thanks also to Kevin Hanson, who arranged publication of this book through Gordon Soules Book Publishers Ltd. The True North Strong and Free Inquiry Society is pleased that Gordon Soules, publisher of the society's first inquiry, The True North Strong and Free?, has accepted the important task of publishing this second inquiry. Soules's confidence in the success of this book is apparent: "I believe the book will make an important contribution to the understanding of the Arctic and its role in international peace and security."

The True North Strong and Free Inquiry Society

Introduction

W. H. HURLBURT

This book is the proceedings of the public inquiry entitled The Arctic: Choices for Peace and Security, which was held in Edmonton on March 18 and 19, 1989, and was the successor of the enormously successful The True North Strong and Free? Inquiry of 1986. Both inquiries were conducted by The True North Strong and Free Inquiry Society, although the 1986 inquiry was sponsored by the Council of Canadians and Canadian Physicians for the Prevention of Nuclear War. The proceedings of the 1986 inquiry were published by Gordon Soules Book Publishers Ltd. under the title *The True North Strong and Free?—Proceedings of a Public Inquiry into Canadian Defence Policy and Nuclear Arms.*

Soviet submarines in Arctic waters can launch missiles that can reach North America. The U.S. navy has devised a forward maritime strategy under which its ships would try to destroy a Soviet Northern Fleet to which those submarines are attached. The threat of nuclear confrontation in the Arctic and the associated threat of nuclear warfare in and over Canadian territory worried many participants at the 1986 inquiry. The Society thought that an inquiry into this complex and dangerous military situation would help both to clarify it and to raise public consciousness about it. And so the 1989 inquiry came about.

The organizers decided, however, that the inquiry should be about more than mere freedom from armed conflict and its consequences, however important that freedom may be. The inquiry was therefore about many things. It was about the needs and aspirations of the native peoples of the Arctic. It was about Canadian sovereignty in the Arctic. It was about resource extraction. It was about the preservation of the environment of the Arctic. It was about the Arctic as a training ground for U.S.

Introduction

and European forces and as a theatre for the contest between the superpowers. It was about the Arctic as a place where the superpowers and other Arctic countries might come together in peaceful cooperation in order to reduce political and military tension. It was about the achievement of military, social, economic and environmental peace and security in the Arctic and in the world.

The 1989 inquiry was conducted in much the same way as the 1986 inquiry. Moderators controlled the discussion, and speakers with different backgrounds and viewpoints discussed the issues and answered questions put to them by a media panel and by members of the audience. There were twenty-four speakers at the 1989 inquiry. The moderators were Stephen Lewis and Adrienne Clarkson; the media panel was composed of Gwynne Dyer, Linda Hughes and Ann Medina.

This book presents the addresses of the speakers and the dialogue between speakers and questioners, with minimal changes made for clarity and stylistic uniformity. Endnotes to the speeches offer citations and references from the speakers, explanatory notes by the editor or short sections of speeches that speakers did not have time to present orally. For any speech where longer sections had to be omitted from the oral presentation because of time constraints, the relevant sections are printed in italics in the body of the speeches.

One theme of the inquiry, and therefore of this book, is the special position of the native peoples of the Arctic, their devotion to their homelands and ways of life, and their wish to control their own destinies. Mary Simon, an Inuk, and (through a message read to the inquiry) Chief Daniel Ishini of the Innu express their passion. That passion is recognized by a northerner, Dennis Patterson, and by three southerners who have at various times immersed themselves in the North: Gordon Robertson, Tom Berger and John Merritt. Tom Berger and Ed Weick discuss particular problems imposed on Arctic residents and the Arctic environment by southern industrialization.

Another theme is international cooperation. Ambassador Esko Rajakoski describes the current efforts of the Finnish government to enlist the Arctic countries in a concerted effort to stop the ruin of the Arctic environment. Yuri Kazmin speaks about the Soviet Union's desire to cooperate in Arctic matters. Walter Slipchenko describes international cooperation in assembling and using information and know-how about the Arctic. This theme merges into that of peace and security in the military sense with Willy Østreng's discussion of the coupling and decoupling of non-military and military issues. It enters into Franklyn Griffiths' plea for civility in international relations.

Yet another theme is sovereignty, particularly Canadian sovereignty, in the Arctic. Canadians become upset at irregular intervals by apparent threats to that sovereignty. These threats include the passage of U.S. ships through the waters of the Canadian Arctic Archipelago without Canadian permission. They include the suspected intrusion into Canadian waters and airspace of foreign submarines and aircraft. Keith Greenaway expresses concern about the effectiveness of Canada's supervision of its airspace. Harriet Critchley points out the need to maintain Canadian sovereignty in order to apply and supply the values that Canadians hold. Gurston Dacks goes beyond sovereignty to advocate a comprehensive approach to Canadian social, environmental and military policies in the Arctic.

The last great theme is peace and security in their military and political aspects. Willy Østreng lays out the strategic background and describes the process by which the Arctic has moved from being a military vacuum to being a military flank and a military theatre. He goes on to describe the later movement towards the decoupling of non-military matters from military security. Johan Jørgen Holst, the Norwegian Minister of Defence, assesses the situation and concludes that the reduction of tensions and confrontation is a global matter and cannot be addressed in isolation.

On the other side of this discussion are Ambassador Alexei Rodionov and Andrey Granovsky. The Soviet Union, they say, wants to negotiate the reduction of tensions both globally and regionally. It wants, they say, to see progress on the Arctic zone of peace proposed in Mr. Gorbachev's Murmansk speech of 1987. It is even willing, they say, to discuss the questions raised by the Soviet Union's immense military and naval installations in the Kola Peninsula and by the Soviet Northern Fleet's Arctic operations. Captain James Bush of the Center for Defense Information in Washington, D.C., and Steve Shallhorn of Greenpeace make somewhat similar points and suggest other ways of reducing tension in the Arctic, as does Gwynne Dyer in his questions.

The speakers and questioners weave a marvellous tapestry in which theme leads to theme and refers back to other themes, all of them divergent and all of them related. The book records these themes. However, it cannot record the rapier-like wit of Stephen Lewis, with which he kept order. It cannot record the gracious presence of Adrienne Clarkson, immersed to the exclusion of any thought of self in maintaining the flow of the inquiry. It does not record the electric atmosphere of the media panel question periods nor much of the passionate desire of questioners from the audience to uncover the roots of the issues. Above all, it does not record the brooding presence of the fifteen hundred people of the audience, who spent the whole weekend before a provincial election attending to every word and nuance of the inquiry.

Was the inquiry a success? The organizers think it was. Could it have been better? Certainly it could.

Introduction

No one from the political level of the government of Canada was willing both to address the inquiry and to answer questions in the inquiry format. In September 1988 Joe Clark, Secretary of State for External Affairs, told us that his department would be represented, but he did not know whether or not he would represent it. By February 1989 other engagements had precluded his attendance. Ambassador Douglas Roche and Major-General David Huddleston, welcome as they were, could not speak for the political level of the government. The participation of Mary Collins, the Associate Minister of Defence, was subject to conditions that precluded questioning by the media panel and the audience. The organizers shared the disappointment expressed by the panel and audience.

The United States government (or at least its executive branch) declined to participate. We received contradictory reasons, but it seems likely that it simply did not choose to appear before a Canadian audience under the inquiry format. The contrast between its reaction and that of the Soviet government in its *glasnost* phase is understandable and was probably predictable, but it is nonetheless regrettable.

The Canadian Petroleum Association said that it could not identify a speaker with the qualifications we suggested. The principal resource extraction corporations declined to participate. The inquiry would have benefited from their direct participation.

However, the marvel was in those who came. Distinguished Canadians lent their expertise, their skills and their names to the inquiry. Distinguished people from foreign countries also did so. It is remarkable that a busy Finnish diplomat would come, and even more remarkable that a Norwegian minister of defence would make a special trip to this country to tell a Canadian audience about his government's views of defence policy. Only a few short years ago it would have been unthinkable for a Soviet diplomat and senior Soviet officials to appear on a Canadian platform, explain their country's policies and answer uncontrolled questions. That is why the inquiry was a success and that is why the reader should find this book fascinating and valuable.

There is, we think, an ongoing process of public enlightenment, of the raising of public consciousness of the issues raised and discussed in the book. People are more and more coming to realize that the Arctic, its peoples, its environment and its resources are important to those of us who live elsewhere. Internationally, governments and peoples are grappling with revolutionary changes in the attitudes of the Eastern and Western blocs of countries. It is our hope and our belief that the inquiry and this book will make a useful contribution to that process, and it is in that hope and belief that I commend the book to the reader.

CHAPTER 1
Opening Remarks

MARY COLLINS

I am pleased, on behalf of the government of Canada, to congratulate The True North Strong and Free Inquiry Society for organizing and promoting this second meeting in its continuing dialogue about Canada, the world and our future.

On a cold and blustery November weekend in 1986, we witnessed an event unique in contemporary Canadian public policy—The True North Strong and Free? inquiry. Five thousand Canadians, from all walks of life and shades of opinion, participated in a public discussion of defence policy and nuclear arms issues. This is the stuff that real democracy is made of.

Your deliberations then, as today, touched upon many of the social and political issues that are before Canadians. I am impressed by the broad coverage of Arctic affairs to be offered by the distinguished panel of speakers that you have gathered here for your inquiry. In particular, I am very pleased to know that my colleague Johan Holst, the Norwegian Minister of Defence, will be joining us for the inquiry.

I am here because of a long-term interest in and commitment to the North, and not as a matter of coincidence or the luck of the draw. In fact, it was fifteen years ago this spring that I flew down the Mackenzie River to Inuvik for the first time. I shall never forget the feelings of that trip, the sense of excitement and awe, as I first viewed those vast lands, the herds of caribou and the twenty-four hour daylight. Thus began a very personal relationship with the North and with northerners. This relationship has shaped many of my attitudes and my beliefs, which are part of my political agenda today.

Over the next five years, as I spent time in the small villages of Tuktoy-

Opening Remarks

Mary Collins

aktuk and Paulatuk and, to a lesser degree, in the Eastern Arctic, I came to know and be part of the people of the North, their hopes and dreams, as well as their concerns. So today I am here not only as a minister of a government committed to the preservation and enhancement of life in the North, but also as an individual, a Canadian, who seeks to be part of a solution and not part of a problem.

Having said this, I recognize that the central issue facing any government is to seek consensus and to find the right balance between competing interests. Prudence and patience are inextricably linked to this delicate balancing act, as we seek solutions to issues where there are no absolute answers.

Those of us whose nations lie around the Arctic Basin must become more involved in and informed about Arctic affairs if we are to make clear judgments about peace and security. The pace of technological, political and climatic developments that affect the Arctic is increasing and must be addressed.

New developments in communications, transportation, resource extraction and military capability have increased the strategic importance of the Arctic. Discoveries about the ozone layer and the greenhouse effect underline the very sensitive environmental role of the Arctic. These developments are drawing together the northern peoples of the world and focusing attention on common interests and opportunities.

It has been said that this is the age of the Arctic, so it is most appropriate that the focus of this inquiry is on the choices for peace and security in that region.

In 1985 the joint parliamentary committee that reviewed Canada's international relations pressed for a northern dimension to Canada's foreign policy. The government responded to these recommendations in 1986 by focusing on the following four broad policy themes:
1. Buttressing our sovereignty over our Arctic waters
2. Modernizing our northern defences
3. Preparing for the commercial use of the Northwest Passage

Mary Collins is Canada's Associate Minister of National Defence. She holds a degree in political science from Queen's University. She pursued a career in the public service of Ontario, becoming the first female executive officer to the Cabinet and Premier. Subsequently she established a communications consulting firm active in industry-government relations and the impact of resource development and industry activities, especially in the North. In 1981 she became director of public affairs for Brinco Ltd. She was first elected to the House of Commons, as the P.C. member for Capilano, in 1984. She won re-election in 1988 and was appointed to Cabinet in January 1989.

4. Expanding our circumpolar relations, including contacts among northerners of different nations

The government's response also stressed the need for consistency in foreign and domestic policy initiatives in order to ensure a comprehensive and coherent Arctic policy.

Canada has been and continues to be active in circumpolar cooperation. Recent agreements have been entered into with Denmark on environmental cooperation and with the Soviet Union for the exchange of Arctic scientists and scientific data.

Additionally, we are supporting the development of the Inuit Circumpolar Conference, which will meet this June. We have intensified our cooperation with Norway on northern issues, which resulted in a bilateral conference held in Tromsø in 1987 led by our respective foreign ministers. In January, some of my parliamentary colleagues attended a multilateral conference hosted by the Soviet Union, which sought new efforts to protect all our Arctic environments through international cooperation.

On the home front, the government is pursuing a domestic agenda that includes both the devolution of provincial-type programs to the governments of the territories and a movement towards an early settlement of native land claims. In pursuing that agenda, focusing on political and economic change and looking for security and prosperity in the North, we must make every effort to preserve the traditional values of our northern peoples.

Limiting excessive militarization of the Arctic in the interests of strategic stability, within the context of our arms control and disarmament efforts, is of particular interest to this government. However, as we pursue these goals, we must temper our idealism with realism. We cannot gamble with our freedom and security—they are too precious.

Over the next two days we will hear a lot about security, freedom and prosperity; they are, after all, the handmaidens of peace. They do not exist in an abstract sense for nations, but are highly dependent on
• the extent to which the rights, values and freedoms of the people and the environment in which they live are protected;
• the economic and social health of the people, individually and collectively;
• the degree of military security enjoyed.

A nation cannot ignore these factors in any of its regions and consider itself truly secure. The multidimensional approach taken by the organizers of this inquiry recognizes these relationships and, I hope, will promote a balanced debate on the choices for peace and security.

For my part, let me say a few words about security, defence and arms control. The security of the Arctic is inseparable from that of Canada as a whole. The threat does not originate in the Arctic, and its elimination does not lie there. It lies in the resolution of East-West tensions.

Let me look for a moment at the East-West relationship, which is unquestionably in a state of flux. Its fundamental nature is changing—we hope for the better.

There is no doubt that the changes we have seen in the Soviet Union—in the field of human rights; in declarations about, and changes in, foreign policy; and in the unilateral commitment to disarmament—are all positive developments. In responding to these changes, we must ask ourselves: How far will they go and how long will they last?

Neither of these questions has a definitive answer. A process that is so volatile can change dramatically in a very short time. Our responses must be crafted to benefit fully from the progress that is made, while protecting us from reversals that could result and threaten our security. This is not an easy challenge and it must be met with imagination and prudence.

I have just returned from Vienna, where I represented Canada at the opening of the new negotiations on conventional armed forces in Europe. I sensed that the assembled ministers from the North Atlantic Treaty Organization (NATO), the Warsaw Pact and the neutral and nonaligned European nations shared an awareness that we have an opportunity now, which we may not have again, to reduce the level of conventional armed forces and confrontation on the fault line of East-West relations.

The results we achieve through these negotiations will set the tone of the East-West relationship for the next generation. Should we fail to act constructively and with patience and determination, the talks could suffer the paralysis experienced during the negotiations of the early eighties. In that event, our security will continue to be threatened by the large imbalances in conventional forces in Europe in favour of the Warsaw Pact.

Never before have the prospects for a mutually beneficial, verifiable agreement been so bright—an agreement that would eliminate the asymmetries in stationed forces and severely reduce the potential for mounting surprise attacks or large-scale offensive operations.

Can you imagine what a different world it would be if the confrontation in Europe were contained and defused? What better impetus for future arms control could there be than the successful completion of a verifiable agreement to this end?

As we stand on the threshold of these negotiations on conventional armed forces in Europe, we can also be optimistic that the Strategic Arms Reduction Talks (START) between the Soviet Union and the United States, aimed at a mutual reduction of fifty percent in strategic systems, will gain momentum as the year unfolds. There is also hope that good progress will be made toward a treaty to ban chemical warfare.

Indeed, this would appear to be an occasion in the course of East-West relations when the interests of both sides coincide. A shared interest exists in reducing the size of armed forces and in siphoning resources from the defence to the civilian sides of our respective economies. We must not let such a chance slip through our fingers. I had the opportunity to convey Canada's desire for progress toward a peaceful, less confrontational world to Mr. Shevardnadze and other foreign ministers in Vienna last week.

As an aside, in my talk with Mr. Shevardnadze, he said, I think neither in jest nor purely in gallantry, that perhaps if all defence ministers were women we would not have any wars.

Over the next two days, as you explore the choices for peace and security in the Arctic, I know that you will approach these issues critically and seriously. I hope that you will agree with me that peace, security and freedom are not alternatives or add-on options—they are integral parts of a whole. An insecure people is not at peace, and peace without freedom is a hollow condition.

We cannot regard the security of our Arctic in isolation from our national security, and we cannot regard Canadian security in isolation from the security of both East and West. As we approach the end of the twentieth century, let us dedicate the decade of the nineties to a renewal of efforts for the achievement of our national objectives: economic security, environmental integrity, tolerance and international cooperation.

Peace, security and freedom are the aims of the government of Canada as surely as they are the aims of all of us here today. Let us all work together to build a lasting peace and let our legacy to future generations be a "True North," truly strong and free.

CHAPTER 2
The Arctic: An Overview

JOHN F. MERRITT

My presentation is entitled "The Arctic: An Overview." The method of presentation I finally struck upon consists of a review of the accuracy of those characteristics popularly attributed to the Arctic.[1]

What then is the popular image of the Arctic for those who neither live there nor have a chance to visit it? All generalizations are suspect, but I think it would be fair to say that many people would believe the following ten things about the Arctic:

- The Arctic is "bloody cold"
- The Arctic is "way up at the top of the world somewhere"
- The Arctic is vast
- The Arctic is almost inaccessible
- The Arctic is barren
- The Arctic is fragile
- The Arctic is unpolluted
- The Arctic is almost uninhabited
- The Arctic is a treasure chest of untapped wealth
- The Arctic is unchanging

Let us examine each of these popular images in turn.

The first is that the Arctic is "bloody cold." Anyone who has been in a place like Resolute Bay in the middle of January with a wind blowing would have no trouble supporting this proposition. It is temperature that determines a commonly accepted southern boundary of the Arctic—the tree line.[2] North of that line summer weather, even in the warmest months, is too cold for coniferous trees to grow successfully. In North America, but not in all parts of the Arctic, the tree line also largely defines the area that has continuous permafrost.[3]

The Arctic: An Overview

John F. Merritt

It is not enough, however, to say that the Arctic is cold without making a few other observations. First of all, there can be tremendous seasonal variations in temperature. Places that have many months of bitterly cold winter experience summers in which temperatures climb into the 10°C to 20°C range. Seasonal fluctuations are particularly pronounced in inland regions. Because of its close integration into the large Eurasian land mass, the Soviet Arctic has very sharp seasonal swings in temperature.[4]

These regional variations in seasonal temperature flux are only part of the picture. While the sea-ice pack that permanently covers most of the Arctic Ocean may have a stabilizing effect on temperature variation, ocean currents and wind patterns create profound temperature differences between various Arctic subregions in the southerly definition of the Arctic itself. Thus the tree line in Canada varies from 69° north latitude near Inuvik to 55° north latitude in northern Ontario. Within the Arctic, the warming effects of the Gulf Stream are massive enough to allow for raising sheep and growing hay in southern Greenland and to keep harbours in northern Norway and at Murmansk in the Soviet Union free from ice year-round. Indeed, it is hard to overstate the significance of the Gulf Stream at the high latitudes; Iceland in some respects would appear to qualify as an "Arctic" country; it is adjacent to the Arctic Circle and is devoid of trees. At the same time, however, the Gulf Stream gives it a mild climate, with mean January temperatures equivalent to those of New York City.

There is another aspect of northern climate worth emphasizing. While most people know that the Arctic is cold, few know that it is for the most part quite dry. Snow and rainfall are sufficiently low for most of the Arctic to be considered a cold desert. Many southern Canadians are under the impression that northern communities are up to their necks in snow; in reality, small amounts of dry, wind-packed snow, combined with frozen lakes and rivers, actually make movement on the land easier in winter than in summer.

The second perception of the Arctic is that it is "way up at the top of

John F. Merritt is executive director of the Canadian Arctic Resources Committee, a public-interest group based in Ottawa. A member of the Law Society of Upper Canada and a barrister and solicitor of the Supreme Court of Ontario, he is a graduate of the University of Ottawa in law and of Carleton University in arts. Mr. Merritt has held the position of legal counsel for both of the national Inuit groups, the Inuit Tapirisat of Canada and the Tungavik Federation of Nunavut. He has been a policy analyst for the federal Department of Indian Affairs and Northern Development (DIAND) and special assistant to David Crombie, then Minister of DIAND.

the world somewhere." This perception, like so many, is partly true and partly false. The relatively low levels of population in the Arctic mean that, for most of us, the Arctic is "somewhere to the north" of us. Yet there are some important qualifiers to this perception that deserve mention.

First, the previously mentioned variability in the location of the tree line means that the Arctic is sometimes found to the east or west as well as to the north. For example, in the Northwest Territories, the tree line— and potentially the political boundary separating two new territories—is more northwest to southeast than east to west.

Second, we tend to be prisoners of the impressions left by the schoolroom maps that first awoke our sense of geography. Any depiction of a spherical object on a flat surface will cause distortion. The standard Mercator projection map of the world leaves the impression that the Arctic is a long thin ribbon at the top of the countries that occupy the high latitudes. Conversely, a polar projection map reveals the Arctic for what it really is: a central basin of water—the Arctic Ocean—surrounded by the islands and coastal regions of the Soviet Union, Alaska, Canada, Greenland and Scandinavia.[5]

There is another point to consider. While most Canadians do not get the opportunity to visit the Arctic, Arctic conditions are in the habit of visiting them. With the exception of those who live in places like the British Columbia coast and Point Pelee, Ontario, Canadians are accustomed to seeing Arctic-like climatic changes for several months every winter. In their dress and in their out-of-doors body language, the people of Winnipeg in January do not appear noticeably different from the people of Iqaluit. This phenomenon has been dealt with by Canadians in a profoundly schizophrenic way. On the one hand, Canada's nordicity—a term coined by Quebec geographer Louis-Edmond Hamelin—and things associated with nordicity, like hockey, Group of Seven paintings, and Bob and Doug Mackenzie, are badges of national identity. At the same time, travel agents do a lively business in February sending people south, and Canadian cities look like they were designed for a southern California climate.

How big is the Arctic? And how accessible? The common assumptions are that the Arctic is both vast and remote.

There is no doubt that the Arctic is a major region of the world, particularly if one takes a somewhat elastic approach to the definition of the Arctic and includes large Subarctic regions in a more loosely defined "North." A few facts serve to illustrate the Arctic's size: the Arctic Ocean is one of the world's great bodies of water; Greenland is the world's largest island; the Soviet Union's Siberia, as that term is used by many geographers, exceeds eleven million square kilometres; nearly forty percent of Canada is above the 60th parallel; and the geographic centre of Canada is near Baker Lake, an Inuit community in the Keewatin region of the Northwest Territories that is considerably north of the tree line. Facts such as these, however, must be kept in perspective. Some Arctic regions, such as the Arctic portions of Finland, Sweden and Norway, are not huge, even by European standards. And compared to oceans like the Pacific and Atlantic, the Arctic Ocean appears quite modest.

Images of the vastness of the Arctic have no doubt been coloured by the traditional inaccessibility of the region. Canadians have been particularly influenced in this regard; after all, the four-hundred-year search for the Northwest Passage played a major role in the political development of the northern half of the continent. The initial perception of the imperviousness of the Arctic—to ships, to roads, to agricultural settlement—later came to serve as a huge buffer, both physical and psychological, against outside meddling in what Canadians came to view as their backyard.

As technology has advanced and distances have shrunk, this cliché of remoteness and vastness has become less and less easy to sustain. That it could not be sustained first became clear in a military sense with the development of northern aviation, beginning with development of northern air routes to Britain in the Second World War. It was driven home by the postwar emphasis on the Arctic as the shortest route for bombers and missiles between the Soviet Union and the United States. Freedom of manoeuvre in the air has been followed by freedom of manoeuvre at sea. We are all aware of the Arctic's accessibility to submarine navigation. Fewer people may be aware that there are already well-developed surface sea lanes along the Soviet Arctic coastline and that there has been talk of commercial traffic some day across the polar ice pack, connecting Far Eastern and European ports.[6]

The isolation of the Arctic has been breached in other ways as well. The introduction of consumer goods and values into aboriginal societies, and the cultural impact of such communications tools as radio, television and computers have forged new ties—some might say shackles—between North and South.

There are three commonly held perceptions of the Arctic that are in many ways closely related: namely, that the Arctic is barren, that the Arctic is fragile and that the Arctic—at least for the time being—is unpolluted.[7]

It is only on closer examination that the biological wealth of the Arctic becomes more apparent to southern eyes. This is not to say that the Arctic can support the same level of biological productivity as warmer climates.[8]

Despite the low levels of productivity, Arctic lands and waters nevertheless produce the renewable resources that have sustained aboriginal inhabitants for thousands of years at a subsistence level and have more recently supported some commercial operations. In the Arctic, land and marine mammals are harvested for their meat and their fur—a reality that has unfortunately put northern peoples and anti-fur groups on a collision course. In both Scandinavia and the Soviet North, reindeer herding has been a traditional economic mainstay and is well developed. The fishing grounds off the Norwegian and Greenland coasts are especially rich— over three million tons of fish are caught in a good year off northern Norway, for example. In recent years, access to northern fish has become a sensitive issue. Most are familiar with the British-Icelandic "cod wars" of a few years back; fewer realize that the future of the Davis Strait fishery played a central role in Greenland's decision to leave the European Economic Community in the early part of this decade.

While the wildlife resources of the Arctic must be used in such a way as to ensure their future abundance, there can be no doubt that sustainable economic development in the Arctic will require exploitation of the animal, fish and bird wealth there. Proposals for multibillion-dollar oil and gas projects may attract big headlines, but the permanent residents of the Arctic are likely to view the lands and seas as both food locker and cultural touchstone for a long time to come.

There is, of course, one other renewable resource in the Arctic that is of considerable value: fresh water. During the 1970s, hydroelectric schemes in northern Quebec and northern Norway sparked confrontation between governments and aboriginal peoples on their respective rights to lands and waters. In more recent years, we have witnessed project proposals involving the massive diversion of water and Subarctic rivers from northerly to southerly flows. In the Soviet Union, these massive diversion schemes appear to have been, at least temporarily, abandoned; Soviet sources suggest that environmental reasons, as well as other factors, have contributed to this result. In Canada, the Grand Canal proposal still lives. This proposal would divert water now flowing into James Bay towards the Great Lakes, with the use of a string of nuclear-powered pumping stations. Its stated purpose is to "fix the plumbing problem of the Great Lakes." Whether or not it would fix the problem is a moot point; it would certainly apply a big plumber's wrench!

Many of you will recall the debate during the most recent federal election between advocates of the government's Free Trade Agreement and a number of environmental groups on whether fresh water would become a commercial product like any other under the agreement. It appears to me that the alarmists have tended to have the better legal argument. Regardless of the fine print, however, it is difficult to escape the conclusion that the continentalist approach to natural resources embodied in the agreement will tend to put greater pressure on Canada to allow open access to Arctic resources generally, be they fossil fuels or fresh water.

The Arctic is popularly perceived as an environmentally fragile area. That perception is well founded. Their low levels of biological, chemical and thermal energy make Arctic ecosystems particularly vulnerable to human disturbance. Plant and animal life operates at the margin of existence, with few metabolic reserves to cope with mechanical or chemical impacts. Everyone is familiar with photographs of the tundra showing truck tracks that are still visible years after the vehicles that made the tracks have come and gone.

This commonly held view of the fragility of northern ecosystems needs only to be qualified in one important respect. Even in an undisturbed state, Arctic species often experience drastic fluctuations in numbers, with local extinction not uncommon. Many a biologist, for example, has been driven to distraction trying to guesstimate the size of a caribou herd, often with disastrous effect on his or her credibility with hunting communities. While northern ecosystems are in general more vulnerable to disturbance than is the case in more temperate zones, biological systems show a remarkable capacity to survive and, given sufficient time, to recover.

While the vulnerability of the Arctic landscape and Arctic waters has been understood for some time, what has become more widely known in recent years is the vulnerability of the entire Arctic to broader environmental changes. For reasons still not well understood, small changes in the global environment often have exaggerated effects in the North. We have all become familiar with the "greenhouse" problems that are likely to result from the buildup of carbon dioxide, methane and other gases through the burning of fossil fuels and the use of chemical fertilizers. Global warming can be expected to have the greatest effects on the high northern latitudes, particularly in the winter—some models have suggested that an increase of 1°C in average global temperatures could mean an increase of 10°C in winter temperatures in the Canadian Arctic Archipelago. The environmental consequences in the North, though hard to predict, could involve increased precipitation, increased river discharge, reduced sea ice, more icebergs and higher sea levels.

There is a tendency on the part of many Canadians to try to look on the bright side of all this greenhouse-effect talk, to think purely in terms of the advantages to be gained in exchanging lined winter boots for toe rubbers in places like Edmonton and Ottawa. Unfortunately, the downside could include such realities as regional famines, the mass movement of peoples and global political upheaval.

The Arctic's environmental fragility means that the common perception of the region as being pristine and unpolluted is, at best, only a partial truth. This is not to say that there are not large areas of the Arctic where the visitor can view a magnificent slice of majestic wilderness offering all the beauty and solitude the travel brochures promise. But the Arctic is no longer immune to the environmental problems that, however perversely, unite the entire world. One doesn't have to dig very deeply to uncover examples to illustrate this point. In the eastern Canadian Arctic, industrial toxins are showing up at shocking levels in the flesh of marine animals and the breast milk of nursing mothers. In Subarctic Quebec, Cree fishermen may have to change their traditional diets because of the high levels of mercury in fish, resulting from large areas of coniferous forest being flooded for hydroelectric projects. In northern Scandinavia, reindeer herds have been condemned as unfit for human consumption due to the radioactive fallout from Chernobyl.[9]

The environmental agenda in the Arctic today must, at one and the same time, focus on two levels. On the one hand, it is necessary to address those relatively localized problems that can impose lasting damage on sensitive ecosystems: examples of what can be done are the removal of polychlorinated biphenyls (PCBs) from around old Distant Early Warning (DEW) Line stations, and the cleanup of oil drums and other debris from exploration sites. On the other hand, it is even more important to tackle the larger problems causing deteriorating global air quality and the poisoning of the planet's oceans. It will take concerted international action, based on a recognition that pollution is no respecter of sovereignty, to deal with these issues. It is only international cooperation, which Canadians should pursue both as individuals and as a people, that will prevent the Arctic Basin sink from becoming the Arctic Basin cesspool.

Another common image of the Arctic is that it is almost uninhabited. Compared to areas such as western Europe and northern India, the Arctic is almost devoid of people. Indeed, even compared to southern Canada, the Arctic is very lightly populated.[10] There are, however, a number of things about Arctic demography that should be kept in mind.

First, while Canada has only 75,000 to 80,000 people living north of 60°, other parts of the Arctic have higher population numbers and densities. The Soviet Arctic is by far the most heavily populated part of the Arctic. Depending on where the line is drawn, some ten million people live in the Soviet North. More than four-fifths of the population of the Arctic are Soviet citizens.

As you might expect, this Soviet Arctic population includes urban centres of some size; Noril'sk, for example, is a nickel-producing city of more than 200,000 people. In considering such figures, it is important to keep in mind the milder climate of the Scandinavian North and the western Soviet North. Canadians are used to thinking of their two territories north of the 60th parallel as the Arctic, but Leningrad sits on the 60th parallel. It is also important to remember that the comparatively large Soviet population in the Arctic is a result both of the development of the extensive oil and gas resources of northwestern Siberia and of an underlying national policy commitment to promoting northern development, which is backed up by the powerful levers of a centrally planned economy.

A second factor to consider about Arctic populations is that a large proportion of residents—a majority in Canada and Greenland—are members of small, distinct, northern indigenous societies. In the West, these societies include about 200,000 indigenous people, including more than 100,000 Inuit and 40,000 Sami, who are also known as Lapps. In the Soviet North, there are perhaps 300,000 people belonging to indigenous minorities, including Chukchis, Yukaghirs, Nenets, Dolgans, Eskimosy and others. While their numbers are not large in absolute terms and are divided among many small villages, usually of fewer than a thousand people, the indigenous peoples of the Arctic have a geopolitical importance that equals their cultural distinctiveness.

In Canada, the negotiation of aboriginal land claims remains incomplete, with agreements having been concluded to cover Arctic Quebec and the Beaufort Sea region, but not the eastern and central Arctic. In other countries, notably Denmark's Greenland with its home rule government, and Arctic Alaska with its North Slope Borough, aboriginal peoples have acquired greater control over their homelands by creating and controlling new political institutions. Canadians used to thinking in terms of Indian reserves in southern Canada might find it difficult to realize that Canada's second-closest neighbour, with a population comparable to one of the smaller members of the United Nations, is run by an Inuit-dominated government within a rather loose Danish sovereignty.[11]

It is also worth noting that the northern indigenous peoples have been adept at forging links among themselves.[12] In the Arctic, the inseparability of domestic and international issues is as relevant to political change as it is to environmental problems.

A final factor to consider about northern populations is that they are growing rapidly. In Canada's North, for example, fertility rates are only now starting to decline from their traditionally high levels; the Inuit population will double over the next generation.

In Western countries, the desire of the central governments to reach political and land claims settlements with northern aboriginal peoples has been largely motivated by a belief that the unextinguished land rights of the indigenous peoples might cause legal headaches in the event of development bonanzas. This notion of the North as an El Dorado is a recurring one. From the Yukon gold rush of the 1890s, to the scramble to build a pipeline from the Beaufort Sea in the 1970s—a contest now being refought through applications to the National Energy Board for gas export licences—the Arctic has been perceived as a treasure chest of untapped mineral wealth.

A great deal could be said about the appropriateness of the treasure chest metaphor in the years ahead—evidence can be amassed both to substantiate and to challenge the notion. Suffice it to say the following:

1. While nobody knows for sure, there probably are large quantities of oil and gas yet to be discovered and tapped in the Arctic, particularly in the offshore areas.
2. The extraction of these hydrocarbons will be expensive, will involve substantial environmental risks and may exacerbate unresolved sovereignty issues, particularly in the Beaufort and Barents Seas.
3. Fast-tracking oil and gas development would compound larger environmental problems. If, for example, Canada is serious about cutting fossil fuel consumption in order to limit a global greenhouse problem, it makes more sense to reduce energy use in the South rather than to secure greater production in the North.
4. The prospects for generating wealth through the extraction of nonrenewable resources should not detract from the ongoing need to promote small-scale, community-based sustainable development for the benefit of permanent residents in the North. Things like tourism are more likely to generate jobs for local people than are megaprojects.

Finally, is the Arctic unchanging?

The answer to this question must be a resounding no. No, because, as previous ice ages reveal and as the greenhouse effect threatens, the Arctic can be a restless place, always expanding and contracting. No, because advances in military and communications technology mean that the Arctic can never be dismissed as impenetrable again. No, because our global environmental dilemma will force us to detoxify or disappear. And finally, no, because the political awakening of indigenous peoples in the Arctic will force southern governments to share control.

Confronted by the inevitability of change in the Arctic, our common challenge is to learn, to adapt and to conserve—and to cooperate in doing so.

Notes

1. The presentation at the inquiry also included slides, each of which, as the speaker observed, was worth a thousand words. (Editor)
2. Across northern Canada, generally speaking, the northern limits of coniferous forests coincide with the mean July 10°C isotherm.
3. Permafrost is ground that retains temperatures below the freezing point of water all year round. It exists wherever the annual mean temperature is at or below that point. This permanently frozen ground is discontinuous at first, particularly where trees or shrubs hold a heavy blanket of insulating snow during the colder months of the year. On the

tundra, permafrost tends to be continuous and, in places, exists to a depth of 300 metres or more.

4. In contrast, the Canadian Arctic Archipelago and Greenland have less spectacular swings due to the moderating influence of surrounding seas that are not continuously frozen.

5. This ocean basin has two openings: the very narrow Bering Strait separating Alaska and Siberia—a strait that once was dry land and allowed successive waves of Asiatic peoples to cross to the Americas—and a broader opening to the North Atlantic, the so-called Greenland–Iceland–United Kingdom gap, which is so important to postwar NATO and Soviet naval strategists. Some scientists have considered the connection between the Arctic Ocean and the Atlantic so significant as to make the former a giant bay of the latter.

6. In 1977, the Soviet nuclear-powered icebreaker *Arktika* became the first surface vessel to reach the North Pole. (Editor)

7. To those of us brought up in urban or temperate landscapes, the North can appear barren. It is easy for me to recall my first impression of the North, when I was left behind by a Boeing 737 in Baker Lake in April; at that time the gravel airstrip had no terminal building. The village was not visible from the airstrip, and as I watched the plane that had brought me disappear into the sky, I felt utterly without reference point: no buildings, no pavement, no farms, no trees. Just a little more humility.

8. The reduced solar energy available for biological processes results in low energy systems. Northern ecosystems, both terrestrial and marine, are characterized by low productivity and generally small populations, typically widely dispersed or migratory, capable of storing energy for long periods, and fluctuating a great deal in numbers. One tropical coral reef, for example, can generate as much biological growth as an immense area of Arctic sea.

9. The Arctic is especially exposed to environmental problems created elsewhere. The Arctic Ocean basin acts as a repository or sink for much of the airborne and seaborne pollution created by the industrial world. Pollutants generated in North America and Europe are washed into the Atlantic Ocean and carried by the Gulf Stream into the Arctic Ocean. The Arctic Ocean ice cover prevents pollutants from reacting with sunlight and the atmosphere; as a consequence, chemical breakdown is minimal. Air pollutants carried into the Arctic from industrial centres in the South result in the problem of Arctic haze. There are major concerns about ozone thinning and ozone holes in the Arctic.

10. These facts are not surprising. With a climate too harsh to support agriculture and with high transportation and energy costs, the Arctic is unlikely ever to become as densely inhabited as most of the world.

11. The inability to distinguish between the circumstances of northern and southern aboriginal peoples has sometimes created major problems in the successful evolution of northern political life. Current federal government land claims policy, for example, seems to reflect the proposition that the Inuit of the Eastern Arctic are doomed to become a minority in their own homeland and that title to a chunk of the land surface can deliver a fair measure of economic self-reliance.

12. Inuit, Crees and Sami have all played high-profile roles in the international recognition of the rights of aboriginal peoples. The northern Quebec Crees and the Inuit Circumpolar Conference enjoy observer status at the United Nations for certain purposes.

Security, Peace and the Native Peoples of the Arctic

Mary Simon

CHAPTER 3
Security, Peace and the Native Peoples of the Arctic

MARY SIMON

The theme of this year's inquiry, namely, "The Arctic: Choices for Peace and Security," is most relevant and timely. It is a question of particular interest to the Inuit Circumpolar Conference (ICC).

The ICC is an international organization whose head office is currently in Canada. Its members are made up of Inuit from Alaska, Canada and Greenland. Since 1983 the ICC has enjoyed nongovernmental organization (NGO) status at the United Nations.

Consistent with the diversity of its charter, the ICC believes that there is a profound relationship between peace, human rights and development. None of these objectives can be realized in isolation from the others. For Inuit in the North, each of these key elements is linked to our environ-

Mary Simon is currently serving a three-year term as president of the Inuit Circumpolar Conference, an organization of Inuit in Alaska, Canada and Greenland, soon to be joined by their cousins in the U.S.S.R. She was educated in northern Quebec and became involved with the delicate political negotiations leading to land claims settlements for native claimant groups in the area, first as a director of the Northern Quebec Inuit Association and later as vice-president, then president, of its successor organization, Makivik Corporation. Ms. Simon joined other Inuit leaders in the successful effort to have aboriginal rights recognized in the Canadian constitution, and has remained actively involved with the Inuit Committee on National Issues. She is a member of a number of boards and advisory committees, vice-chairman of the board of the Native Economic Development Program, and a director of the Canadian Institute for International Peace and Security.

ment and to the lands, waters, sea ice and resources upon which we as a distinct people depend. Any measurement of security in the Arctic must take into account all of these factors.

The ICC is working towards reversing the trend to militarize the North. We are committed to promoting peaceful and safe uses of the Arctic. We strive for initiatives that benefit not only those of us whose home is in the North, but all of humankind.

The ICC firmly believes that extensive circumpolar cooperation (for example, international trade, polar research, environmental protection, cultural exchange) would be a major step toward promoting Arctic and global peace. However, such cooperation would flourish best if it included meaningful and comprehensive arms control initiatives.

It is in this spirit that the ICC has been working for a number of years towards the development of a comprehensive and coherent Arctic policy. Basic principles and elements will be completed soon on a wide range of economic, social, cultural and political issues, including environmental and defence-related matters. Our latest policy work will be submitted by Inuit delegates for review at the July 1989 ICC General Assembly in Sisimiut,[1] Greenland.

In addition, the ICC continues to develop a transnational Inuit Regional Conservation Strategy (IRCS) for the Arctic. For its ongoing environmental work, the ICC was honoured as a 1988 recipient of the Global 500 Award by the United Nations Environment Program.

The subject that I was asked to address today is "Security, Peace and the Native Peoples of the Arctic." As ICC President and as an Inuk from northern Quebec, I will be speaking from an Inuit perspective.

The thrust of my presentation is that choices for peace and security in the Arctic are being limited unnecessarily by states in the Arctic, to the detriment of our common objectives. Moreover, if Inuit and other aboriginal peoples in the Arctic continue to be effectively excluded from policy- and decision-making processes relevant to the North, it is highly unlikely that lasting peace and real security—as we perceive it—will be achieved.

In discussing Arctic peace and security issues, I will focus on the following aspects: (1) current perspectives on peace and security options in the Arctic that we feel are too limited; and (2) some steps that we propose to advance peace and security in the Arctic.

I will first discuss the limited perspectives on peace and security in the Arctic.

In addressing problems related to the Arctic, it is vital to recognize that vast regions in northern Canada, Alaska, Greenland and eastern Siberia constitute, first and foremost, the Inuit homeland. We do not wish our traditional territories to be treated as a strategic military and combat zone between Eastern and Western alliances. For thousands of years, Inuit have used and continue to use the lands, waters and sea ice in circumpolar regions. As aboriginal people, we are the Arctic's legitimate spokespersons.

Since our northern lands and communities transcend the boundaries of four countries, we are in a unique position to promote peace, security and arms control objectives among states of the Arctic Rim. Any excessive military buildup in the North, whether by the Soviet Union or the United States, only serves to divide the Arctic, perpetuate East-West tensions and the arms race, and put our people on opposing sides. For these and other reasons, militarization of the Arctic is not in the interests of Inuit who live in Canada, the Soviet Union, Alaska and Greenland. Nor do such military preparations further security or world peace.

The ICC is encouraged by the signing of the Intermediate-range Nuclear Forces (INF) Treaty in December 1987. However, we must be vigilant that the removal of an entire class of nuclear weapons in Europe does not give way to an arms buildup in other regions of the world, such as the Arctic or the North Pacific.

In terms of the North, Inuit seek to decrease significantly the possibility of armed conflict, whether it occurs by design or by accident. We believe that demilitarization of the Arctic in a gradual, balanced and fair manner is the most productive course to pursue at this time.

For defence, environmental and other matters, governments of Arctic states must look beyond geographical and political borders and alliances. Like the Inuit, they must begin to perceive and value the Arctic as an integral whole that not only sustains our Inuit way of life but also determines the earth's future in many ways. It is in this context that Inuit support the idea of working towards the formal establishment of the Arctic as a zone of peace. Consistent with such a zone must be the reduction and eventual elimination of nuclear weapons in ever-increasing parts of the North.

To date, there is little evidence that most states in the Arctic are prepared to seriously consider new approaches or options for security and peace in circumpolar regions. Our present view finds support in the following three examples.

First, little or no effort has been made to replace Cold War military doctrine with more forward-looking strategies and concepts. For example, the "threat" facing non-nuclear-weapons states in the North should be more accurately described in terms of a possible nuclear conflict between the superpowers and not just in terms of a nuclear attack by the Soviet Union. This necessary adjustment in perspective could lead to significantly altered responses to potential nuclear threats in the Arctic.

Second, the notion of security should be considered in broad terms of collective security for all peoples and all countries. States in the Arctic must not continue to narrow the discussion to just national, continental or Western security in military terms.

In order to strengthen security in the North, increased militarization is hardly the path to a lasting solution. Immediate steps must be taken cooperatively by nations in the Arctic and by other nations to counter transboundary pollution effectively. Polychlorinated biphenyls (PCBs) and other persistent chemicals are seriously jeopardizing the health of Inuit, our northern environment and our wildlife. In addition, the Arctic and other regions of the world are being threatened by the continuing destruction of the ozone layer by chlorofluorocarbons (CFCs). While some positive steps are being taken by governments, increased measures are still needed. At this point, threats to the Arctic and global environments, not military threats, pose the greatest dangers to the security of the North and other regions of the world.

Finally, non-nuclear-weapons states in the North do not appear to be determining their own priorities and policies in regard to arms control based on the values, interests and concerns of their own respective countries. In particular, strategies and positions being put forward by these countries are not the sort that would advance peace and security objectives in the Arctic.

The ICC believes that concerted pressure must be brought to bear consistently on both the Soviet Union and the United States by non-nuclear states, if a concentration of weapons systems in the Arctic is to be prevented.

In October 1988 Canada's Secretary of State for External Affairs, Joe Clark, outlined his government's view on the Arctic, in a speech at Carleton University in Ottawa. Mr. Clark indicated that allies in the North Atlantic Treaty Organization (NATO), including Denmark and Norway, agree that security in the Arctic cannot be dealt with in isolation—it is a NATO issue and not a northern issue. The ICC accepts that Arctic security need not be addressed in isolation but believes that Arctic-specific measures may still be needed in order to avoid a weapons buildup in the circumpolar regions. To describe security in the Arctic as a NATO issue and not a northern issue restricts this crucial notion to military terms. In addition, it minimizes the role and responsibilities of Canada as an independent and sovereign Arctic state with its own particular northern interests. Equally importantly, it serves to ignore the rights, concerns and priorities of Inuit in the Arctic and to deprive us of a meaningful role.

I now turn to steps that the ICC proposes for advancing peace and security in the Arctic.

Without effective and ongoing Inuit involvement, we feel that the full range of Arctic concerns and options for peace and security will likely be neither identified nor appropriately addressed. I would like to share with you some positive actions that could be taken to further peace and security in the Arctic. We feel that there is a wide range of measures that could be taken by Arctic states and others to advance security and peace objectives in circumpolar regions. In some instances, the steps we propose are Arctic-specific, while others are of a more global nature.

These measures include the following:

1. The need to promote internationally the illegality of nuclear weapons and to seek constructive, non-nuclear deterrents on which security might be based.

2. The need to redefine the notion of security in broad terms of collective security for all peoples and states.

3. The need to undertake seriously and to support research efforts towards the establishment of the Arctic as a zone of peace. In this context, the possibility of creating nuclear-weapons-free zones in significant areas of the Arctic (in conjunction with other measures) should be thoroughly and openly examined.

4. The need to respect the fundamental values and rights of Inuit and other aboriginal peoples in the Arctic. A primary and explicit objective of Arctic policy (which is currently lacking) must be to ensure the security and well-being of aboriginal peoples in ways acceptable to them.

5. The need to incorporate impartial and timely environmental and social impact assessment procedures in all aspects of arms control and defence planning for the North. This is especially important, considering the global interests in safeguarding the integrity of the Arctic environment and the profound significance of northern lands, waters and resources to aboriginal peoples in the Arctic.

6. The need to ensure the direct and ongoing involvement of the Arctic's aboriginal peoples in policy- and decision-making on all issues, including those related to arms control and defence.

7. The need to include in an evolving Arctic policy framework such emerging human rights as the right to peace, the right to development and the right to a safe and healthy environment. In early August 1988 the ICC emphasized these emerging human rights in Geneva at the 6th Session of the United Nations Working Group on Indigenous Peoples. Unfortunately, the Arctic state government representatives present did not voice support for these vital concepts.

8. The need to elaborate sufficiently the relationship between military spending and social development in an Arctic context. Although northern economic opportunities are critically needed, militarization should not replace proper socioeconomic development in the Arctic.

9. The need to assess fully the impact of the new technologies on Arctic and global security. In this context, the recent approval to test advanced cruise missiles over the Canadian Arctic should be re-examined.

10. The need for non-nuclear-weapons states in the North to identify key issues of concern fully (for example, cruise missiles, prohibited zones for armed submarines) and to pursue negotiations for arms limitations within Arctic regions.

While military activities continue to be justified by governments on the basis of defence and security considerations, such actions often serve to promote our insecurity and may threaten our unique and essential Arctic environment. These activities may also conflict with aboriginal use of circumpolar lands, waters and sea ice.

Despite repeated requests, in-depth involvement of Inuit in northern defence and arms control matters has yet to be accommodated. If the future of our Arctic homeland is to be safeguarded, we must have direct and ongoing input in government policy-making. Meaningful Inuit participation could contribute to a vital form of circumpolar cooperation that should not be ignored.

We are convinced that real security and peace in the Arctic, and elsewhere, rely to a large degree on political, not military, solutions to East-West rivalry and tensions. To date, most states in the Arctic have been less than energetic or innovative in formulating arms control proposals and in responding to initiatives from opposing states.

In the North, our immediate security is being threatened by persistent and severe environmental pollution that must be cooperatively addressed by all nations. We are aware that national budgetary deficits are currently a serious preoccupation of most Arctic countries. However, we urge that precious human and financial resources be reserved by governments on a priority basis for comprehensive and ongoing environmental action.

Inuit are hopeful that Arctic Rim nations will not remain submerged in the policies of the past. Now is the time to shape new visions and prepare fresh blueprints for circumpolar cooperation. Collectively, we can breathe new life into the concept of security. Only together can we ensure that the Arctic will be in the future what it was to our Inuit ancestors—a land of sustenance and a land of peace.

Notes

1. The Greenlandic name for Holsteinsborg, West Greenland. (Editor)

CHAPTER 4
The North as Frontier and Homeland

THOMAS R. BERGER

In 1977 I completed my work in the Mackenzie Valley Pipeline Inquiry and wrote a report called *Northern Frontier, Northern Homeland*. An editorial in *The Globe and Mail* this morning said the time had come to consider the future of the Mackenzie Valley in the post-Berger era. Now that I'm no longer a judge or a royal commissioner, but just a private citizen in Vancouver, may I be the first to weigh in with views on the subject.

Next month, the National Energy Board is going to hold hearings on proposals for exporting gas by pipeline from the Mackenzie Delta along the Mackenzie Valley and then south to the metropolitan centres of the United States. I considered a proposal for such a pipeline in 1977 and I urged that it be rejected, but not on environmental grounds. I felt then, in 1977, that there were no overriding environmental reasons for rejecting a gas pipeline along the Mackenzie Valley. I urged that there be no pipeline for ten years to enable native land claims to be settled, to enable the subsistence economy—the traditional economy based on hunting, fishing and trapping—to be strengthened and to enable the concept of sustainable development to guide the future of the North. Indeed, I urged in 1977 that we think of development not just in terms of large-scale, capital-intensive frontier projects, but also in terms of strengthening the traditional economy on which indigenous peoples throughout the Arctic and Subarctic have depended for hundreds and hundreds of years.

Now twelve years have passed. The claims of the Inuvialuit in the Mackenzie Delta were settled in 1984. An agreement in principle was reached between the Dene and the Métis and the government of Canada last summer and was signed by the prime minister of Canada and the leaders of the Dene and the Métis at Fort Rae. Therefore, the conditions

The North as Frontier and Homeland

Thomas R. Berger

that I felt had to be met before a pipeline could be considered have, to a large extent, been met. It is logical that the industry should renew proposals to build a Mackenzie Valley pipeline, and I am content to leave it to the National Energy Board, to northerners, to the governments of the Northwest Territories and Yukon and last, but not least, to the government of Canada to make their own choices about such matters.

May I, having said that, turn to make the point that a Mackenzie Valley pipeline, even if one were to be authorized following the gas export hearings and then following an application to build a pipeline, would not be constructed until the mid-1990s perhaps, at the earliest. That brings me to a subject that I think is more immediate than the Mackenzie Valley pipeline proposal, and that is the future of the Arctic Coast.

When I conducted the pipeline inquiry ten years ago, I had to consider the future of the Porcupine caribou herd, whose calving grounds are found along the North Slope, about a hundred miles to each side of the international boundary between Alaska and Yukon. Based on the evidence that I heard at the time, I recommended to the government of Canada that there should never be a pipeline built or an energy corridor established along the Arctic Coast of Yukon or along the Arctic Coast of the contiguous area of northeastern Alaska.

That two-hundred-mile area along the coast on each side of the international boundary is where the Porcupine herd has its calving grounds, where the herd arrives in June each year like clockwork and where the calves are dropped and where the herd remains until the calves are able to move with the herd. In the fall, the herd moves back into the mountains. It's a herd of 150,000 animals, one of the last great caribou herds in North America.[1] It is a herd upon which approximately ten Indian and Inuit villages in Yukon, the Northwest Territories and Alaska depend. They depend upon it for their traditional food, which constitutes a very large portion of the diet of those northern peoples even today. When you see

Thomas R. Berger currently practises law in Vancouver, where he began his legal career after studies at the University of British Columbia. He was active in politics in the 1960s, serving as member of parliament and member of the B.C. legislative assembly for the NDP. He was appointed to the B.C. Supreme Court in 1971 and resigned in 1983 in order to be able to speak freely. He has headed royal commissions of enquiry, under NDP, Liberal and Conservative governments, on family and children's law in B.C., on the social, environmental and economic impacts of a proposed Arctic gas pipeline in the Mackenzie Valley, and on Inuit and Indian health care. He also headed the Alaska Native Land Claims Review Commission. He holds honorary degrees from thirteen universities and has lectured widely in both Canada and the United States.

that herd moving into the mountains in the fall—150,000 animals—it is one of the great sights on earth. It is comparable to the migration of the wildebeest on the Serengeti Plain in East Africa. We Canadians are the joint custodians with the Americans of this herd, not only for the Inuit and the Indians on both sides of the border, who depend on it, but for future generations.

In 1977 I urged that we in Canada establish a wilderness park in northern Yukon, and Canada has done that. I urged that a wilderness area be established on the American side. The recommendation I made was that the Arctic National Wildlife Range in the contiguous region on the Alaska side be designated wilderness, and I went down to Washington and testified before the Senate and House committees urging that this be done. President Carter adopted that proposal and Congress acted on it in 1980.

Under President Reagan, however, the U.S. administration proposed that the entire Arctic Coast be opened up to oil and gas exploration and development. The administration urged that industrial man should occupy the calving grounds of the Porcupine herd. Congress resisted that proposal throughout the Reagan years, but when President Bush spoke to Congress earlier this year on television, he specifically urged that the Arctic National Wildlife Range be opened up to oil and gas exploration and production. If this occurs, the future of the herd will be threatened.

We cannot protect the Porcupine herd except by joint Canadian and American action. The herd is an international resource. It moves back and forth across the international boundary, unaware of the mandate of national sovereignty that we and the Americans enjoy. It is essential that we in Canada insist that the Americans protect the calving grounds of the herd on their side of the border.

If the area that President Bush wants to open up were opened up and if oil were found, it would, according to the most optimistic projections, keep the United States supplied with oil for two years. That is what we are weighing against the future of the Porcupine herd, a legacy that we are obliged to protect for all mankind and for the people there who depend upon it. I urge that Canadian environmental groups support the government of Canada in seeking to ensure that the Americans maintain the regime of international cooperation established in the early 1980s, which has so far protected the herd.

I recently wrote to Mr. Clark, the Secretary of State for External Affairs, on this subject. I will read his reply. I am sure that he will not mind, and I think the letter does great credit to him. He thanked me for my letter and he said, "When I met with United States Secretary of State James Baker on February 10, 1989, I told him that we believe, we Canadians believe, the most effective way to protect the Porcupine caribou herd and other shared wildlife from the effects of hydrocarbon activity in the Arctic

National Wildlife Range would be for the United States to give the coastal plain of the refuge full wilderness designation." I hope that all of you will support Mr. Clark in that endeavour.

The fate of the Porcupine herd illustrates what really is entailed in the fate of the Arctic. We can protect the Porcupine herd if two countries, Canada and the United States, work together, and we can protect the subsistence economy on which those ten villages on both sides of the border depend. We can, if we are willing to redefine the whole idea of progress, come to the realization that the subsistence economy of hunting, fishing and trapping based on renewable resources is exactly the kind of sustainable development that the Brundtland commission[2] had in mind. That is a concept that can be applied throughout the Arctic, so that the people for whom Mary Simon spoke this morning can be assured that the resources that they have depended upon for centuries will not be lost simply to satisfy the national ambitions of the states in the region.

In my 1977 report I made a number of recommendations; most were adopted, so I can't complain. I now raise another recommendation that I made then, only because I think it is still current and important. I believe it is one that Ambassador Rajakoski intends to discuss. It is an idea common throughout the Arctic; it is not an idea that originated with me. In 1977 I recommended that Canada take the lead in urging the establishment of a circumpolar program of research to determine the impact of industrial development on Arctic waters and Arctic marine life throughout the whole circumpolar basin. I said then, and I repeat it now, that we should consider the impact of offshore oil and gas exploration and production activity. The Americans are proceeding offshore. The Canadians proceeded offshore in the late 1970s. The Danes drilled off the western coast of Greenland in the 1970s. The Soviets have an immense Arctic coastline and they are proceeding offshore. We must consider what the impact would be if all of these nation-states over the next generation were to engage in oil and gas exploration and development throughout the circumpolar basin.[3] We know that even in the temperate zones of the earth we cannot clean up oil spills. We know that in the Arctic, if there were oil spills and oil blowouts over a period of time throughout the circumpolar basin, there would be an accumulation of oil under the ice over a period of years. We don't yet know what the impact would be on Arctic weather systems and Arctic marine life.

I urge then that there be a circumpolar program of research, under a circumpolar treaty, to ensure that all of us, all those nation-states who are the joint custodians of the circumpolar basin for all mankind, participate in the stewardship of the resources of that region. I also urge that the impact of the accumulation of Arctic haze, about which U.S. scientists now know a great deal, should be considered. I also urge that the impact be

The North as Frontier and Homeland

considered of such projects as the Soviet Union's proposal, made in the late 1970s, to reverse the flow of six great rivers leading into the Arctic Ocean—to dam them and to send the water back south to Central Asia to irrigate the deserts there. This proposal was opposed by Soviet environmentalists, but it was not scrapped until Gorbachev succeeded the *ancien régime*. All of these kinds of projects must be carefully considered because capitalists and Marxists have the same devotion to material progress, the same devotion to large-scale, capital-intensive frontier projects.

If you look at the Arctic Ocean on a circumpolar map, you will see that it is really an international lake. All of us have an obligation to ensure that this international lake is protected. I suggest that we should not, in Canada, engage in noisy assertions of sovereignty. Sovereignty should not define our policy. I urge instead that we take the lead in introducing proposals for circumpolar cooperation.

In the North it would be a mistake to permit the concept of national sovereignty to define the terms of the debate. Why should we allow the sterile goals of the nation-state to define the future of the North? National sovereignty is a limited and limiting concept. Beyond sovereignty lies stewardship. Sovereignty is a national issue, stewardship an international issue. Surely the Arctic is a place where we ought to attempt to transcend the particularities of the Cold War.

I just want to conclude by reading from my report. It was written in 1977, but a Vancouver publisher, Douglas & McIntyre, brought out a revised edition just a couple of months ago and I wrote a new introduction to it. I will read a passage from that introduction as the conclusion to my speech today. May I take the liberty of mentioning this new edition of *Northern Frontier, Northern Homeland* and urge you to read it if you think it will be useful in the struggle to sustain the values we cherish in the Arctic. Please remember that I receive no royalties; although the publication of the book assuages my vanity, it does nothing for my pocketbook. So let me conclude by saying:

> Our notions of progress have acquired a technological and industrial definition. For many, the advance of industry and technology to the margins of the globe represents a kind of manifest destiny for industrial man. For others, it represents an unacceptable threat to the future of the biosphere itself. I wish to avoid being thought of as a partisan of either view. But I do urge [a treaty encompassing an international program of research into the impact of industrial development throughout the circumpolar basin on arctic waters, arctic marine life and arctic weather systems][4] so that a complete examination of the impact of industrial activities on circumpolar waters can take place and so that the advance of industry and technology in the arctic and subarctic seas will take

place in a rational and orderly fashion, so that the circumpolar basin—the heritage of all mankind—will be protected.

It is in the North [I believe, speaking to you as Canadians][5] that the survival of the native subsistence economy is essential; it is there that the place of native peoples within our political system will be determined; it is there that our commitment to environmental goals and international co-operation will be tested. In the North lies the future of Canada.

Notes

1. According to a recent compilation, there were then some 2.3 to 2.8 million caribou in North America, in 102 herds. The seven largest herds were George River (600,000), Bathurst (320,000–450,000), Beverly (250,000–420,000), Western Arctic (over 200,000), Kaminuriak (180,000–200,000), Porcupine (150,000) and Northeastern Mainland (110,000–130,000). See T. M. Williams and D. C. Heard, "World Status of Wild *Rangifer tarandus* Populations," *Rangifer,* Special Issue No. 1 (1986): 19—28. Harstad, Norway: Nordic Council for Reindeer Research. (Editor)

2. The World Commission on Environment and Development was created by a resolution of the 38th General Assembly of the United Nations in the fall of 1983, and produced its report, *Our Common Future,* in the spring of 1987 (Oxford: Oxford University Press, 1987). The Commission was chaired by Gro Harlem Brundtland. Among other recommendations, the Commission called for an era of "sustainable development," driven less by short-term economic rewards and more by concern for the long-term welfare of humanity. (Editor)

3. On March 24, 1989, six days after the inquiry began, the supertanker *Exxon Valdez,* carrying Prudhoe Bay crude oil, hit a reef in Prince William Sound, Alaska. An estimated forty million litres of oil, the largest North American spill to date, gushed into the sound, which teems with sea birds, sea mammals, fish and other marine organisms. (Editor)

4. This information was not read by Mr. Berger; it has been included for clarity. (Editor)

5. Mr. Berger added these words in his speech at the inquiry. (Editor)

Responsible Development and Associated Considerations

Edward R. Weick

CHAPTER 5
Responsible Development and Associated Economic, Environmental and Social Considerations

EDWARD R. WEICK

I will try to cover several facets of resource development, and I want to preface my remarks by stating that I am neither in favour of major resource development in the North nor in support of it. But I do think that, at this time, we have to recognize that there is a drift favouring the inevitability of northern resource development. You pick up this drift in a variety of ways. The price of oil is stabilizing, perhaps at prices high enough to encourage northern activity. In Calgary there is a sort of pessimistic optimism brewing. The various oil and gas companies are beginning to become more active. There is the gas export application before the National Energy Board, eyes being, of course, on the continental energy market as opposed to the southern Canadian market. There are pipeline plans—Foothills has come back rather strategically; Polar Gas is still on the table. There is a move in the direction of development, and I do think that one has to recognize it and begin to deal with it in a pragmatic way.

Edward R. Weick is an Ottawa-based consultant providing advice to special inquiries, corporations and governmental task forces and departments. He is an economist with degrees from the universities of British Columbia and Ottawa. Prior to 1972, he held positions with the federal departments of Transport, Defence Production, and Indian Affairs and Northern Development. Since then, he has contributed to the work of the task force on Northern Employment and Economic Opportunities, another task force examining northern pipeline proposals, and the Mackenzie Valley Pipeline Inquiry. More recently, he worked again with government and with Dome Petroleum, before establishing his consulting practice.

A first point is that oil, gas and mineral development may be the only available means of indigenously generating much of the wealth that is needed in northern regions to provide jobs, infrastructure, housing, education and social services at currently acceptable levels. The Arctic, beautiful as it is, pristine as it is, does not generate very much wealth and can probably only do so through major resource development.

A second point is that northern peoples have now made a commitment to modern industrial development. Nobody is wholeheartedly committed to development of this kind, but, to some extent, people have voted, if not with their feet, then with their signature pens. They have signed land claims agreements and agreements in principle; they have, via their governments, the Yukon government and the government of the Northwest Territories, given support to major resource development. In other words, they have come, certainly since the 1970s, to recognize the import of what major resource development could mean to their economic well-being and through that to their social well-being.

I would like to make the point, however, that this support is always qualified. There is a recognition that major resource development is not a panacea. Tom Berger once remarked that to dine on oil and gas is no free lunch, and I think that is still very much the case. If that wasn't apparent before, we certainly learned it in the early 1980s.

A final point, which is probably obvious and which has been made by many other people, is that resource development like other forms of development is an extension to the North from the South. It will happen because we in the South want to continue to switch on our lights. It will take place because we want to drive our cars and use our air conditioners. As idealists we may deplore the industrialization of the North, but as pragmatists we support it. Frontier development must therefore take place. It's not benign. No form of development is benign, environmentally or socially. You have the consequences to deal with as well as the benefits.

There is an urgency to the development of northern resources. John Merritt mentioned that the northern population is growing very rapidly. It continues to grow very rapidly even though there has been a fall in the natural rate of increase. We could well see a doubling of the Canadian northern population within a period of twenty or thirty years. This is probably true of other circumpolar regions as well. Recently a manuscript report entitled *Lords of the Arctic: Wards of the State,* written by Colin Irwin[1] and based on research at Chesterfield Inlet, described a state of drift and failure among the Inuit that is really quite difficult to imagine and quite difficult to accept. The institutions that were to prepare young Inuit for life in the late twentieth century appear to have failed to do so. I'm basing what I am saying on Dr. Irwin's research, which concludes that the next generation may very well live in what he describes as "arctic ghettos." There is, of course, a land-based economy that has consistently given a yield, and this will continue. In 1988 or 1989 dollars it has consistently yielded from $50 million to $75 million worth of produce in the Northwest Territories alone. But it can't support everybody anymore. People will have to leave the villages. People will have to find something to do, perhaps elsewhere. There is that kind of urgency to the need for development.

I don't know whether that urgency really has to be met by major resource development; that is, whether there isn't some other way of providing the kind of money needed to develop the infrastructure, education and services that the people want very much (just as we want them here), but I can't think of any other way. There really does not seem to be any other source of major wealth, of income, of jobs at the level that seems to be needed.

The North is a major deficit region on a national fiscal basis. Much of the development that one sees there now is a result of direct federal expenditures and transfers from the federal to the territorial governments. Between 1974 and 1984 federal expenditures in the territories grew from $200 million to $1.7 billion. Admittedly, Petroleum Incentives Program (PIP) grants are included in these figures. But the growth, even if put in constant dollars, is quite startling. Meanwhile, the amount withdrawn in taxes and various other flows to the national economy from the North didn't increase commensurately. So you have a widening deficit, a widening inability of the North, in a fiscal sense, to pay its own way.

In saying that, I do not intend criticism. I intend simply to state a fact. While the fiscal dependence of the territories is considerably greater than that of the poorest provinces, it may not be as great as fiscal dependence in some economic sectors. A lot of us are subsidized. We don't recognize or acknowledge it but we are. Farmers are subsidized, fishermen are subsidized, and so on. Farmers in fact, on a per capita basis, may now be more heavily subsidized than northerners. The difference is that the political clout that groups like farmers have and the political clout that groups like northerners have may be quite disproportionate; I think that the Meech Lake Accord is perhaps testimony to that. Increasingly, given a tightening fiscal regime in Ottawa, given the deficit, given the national debt and all of these things we can't somehow seem to cope with all that well, the willingness of the government to subsidize regional populations is inevitably going to decrease, with northerners at the end of the queue.

Although the fiscal situation does not in itself justify major resource development, it is an important reason for it. The territorial governments have now signed agreements in principle (the Northern Accords)[2] with the government of Canada. A provision of these is that, when oil and gas development come, a proportion of the royalties—though the amount has still to be negotiated—will be siphoned off and will provide an increasing revenue base for the territorial governments and native claimant groups.

What can major resource development mean for a region? The North Slope of Alaska is a case in point. It has a native-controlled regional government, the North Slope Borough. During the 1970s a capital improvements program was implemented by the North Slope Borough, and it made significant inroads into the poverty and isolation under which the Alaskan Inupiat lived. The infrastructure and services put in place required large sources of revenue, and these came from the taxation of the property of oil companies operating on the North Slope. That property base was valued at $10 billion in the early 1980s; quite a significant taxation base!

Very, very few Inupiat actually worked for the oil companies, as there was plenty of work in the borough. The people didn't have to go and work in the oil patch. They could work on construction, in Barrow and in the other communities, and at a variety of other things. What the borough was able to accomplish was impressive. Prior to the capital improvements program, the typical Inupiat household consisted of five or six people occupying drafty quarters averaging 630 square feet, which may be reminiscent of parts of our own North right now. Honey buckets served as toilets. Surface water of highly questionable quality was used for drinking. Median family income was only about seven thousand dollars a year and it made very little sense to stock grocery store shelves with anything but the most basic staples. If Inupiat youths wanted to attend high school, they had to live with relatives in Fairbanks or, more commonly, attend school in Sitka, some 1400 air miles from Barrow. As a result, in 1970 the average Inupiat adult had only 5.6 years of education.

In the decade following 1974, more than four hundred housing units were built, almost forty percent of the North Slope housing stock. Average household size for Inupiat families dropped markedly by 1977. Houses became much larger. New facilities built or planned for virtually all North Slope communities included schools, electric generation and distribution systems, and water supply and sewage disposal systems. What I want you to understand from these statistics is that the Inupiat did a lot of things themselves with the revenues they brought in. Unfortunately, however, they got themselves into rather heavy debt as well, and that is the downside of the lesson.

I would argue that, by a variety of means, northern peoples have accepted the prospect of major resource development and indeed have committed themselves to it. I recently visited the offices of Northern Transportation Company Limited (NTCL). That company is now owned by the Inuvialuit Development Corporation (IDC) and the Nunasi Corporation. The executive offices are most impressive. Recently NTCL has not made very much money because of the downside of the oil boom, but it expects things to pick up again soon.

Given that native people have put money into major northern industry, they are vulnerable to the cycles that plague such industries. The acceptance of development, like everything else, has two sides to it. They have to share in the risks and losses as well as the profits. Not only that, but the kinds of conflicts that used to distress people, such as the conflicts between major resource developers and native people and the conflicts of the 1970s between pipelines and land claims, have now been moved into the native camp. Perhaps that is where such conflicts belong. These are important questions, and I'm not saying we shouldn't worry about them, but it is the native people who have to make the decisions. They have to decide how much development, how much subsistence, how much commercial exploitation of nonrenewable resources there should be. They have been put in that position via the claims agreements and via buying enterprises like NTCL. They have been put in that position because, via buying NTCL and via buying into drilling for oil and helping to develop the Norman Wells oilfield, they have committed themselves to major resource development. That is not a sellout; it's an accommodation. It's getting on the bandwagon but still trying to preserve what you have. There are still trade-offs, but it's the native people who have to make them now.

It is not only at the organizational level that native people need cash. They also need cash as individuals. The picture of aboriginal hunters that existed some ten to fifteen years ago is no longer an accurate picture. They now hunt much as modern farmers farm. They hunt with snowmobiles, high-powered rifles, even aircraft. All this costs money. In a recent study I estimated that it may cost an individual hunter ten to twenty thousand dollars a year to hunt. Where is that money going to come from? It can't very well come from welfare; welfare doesn't really provide for that. Territorial government programs help. The fur industry is not too helpful; sealing is not too helpful. The main source of cash for the native economy has become wage employment with government or with industry.

The downside of all of this is that there are busts as well as booms, and certainly the North Slope Borough has experienced both. The borough's debt load, when I last heard, was close to a billion dollars. For four thousand people that is a rather high debt. The boom-bust problem is well illustrated by Alaska. In Alaska, oil and gas activity led to unprecedented growth between 1973 and 1982. State government spending was the principal medium for transmitting the impetus provided by the oil and gas industry to the economy as a whole. In 1969 the state received a $900 million bonus for North Slope lease sales on state lands. Regular oil and gas revenues increased sharply during the 1970s. In 1972 the state received $47 million in revenues, or about eighteen percent of its total revenues, from petroleum-related sources. By 1982 this had increased to $4 billion, or eighty percent of the total revenues required to finance a greatly increased state budget.

Alaskans were faced with choosing how much of their oil revenues should be spent and how much should be saved. They appear to have opted for spending most of the revenues and saving relatively few of them. To make matters worse, money was not spent on items that would maintain the revenue-generating capacity of the state after revenues from petroleum began to decline. Other steps taken by the state government also suggest that it was not overly concerned about future revenue sources. In 1980 it abolished the state income tax. In the same year, it enacted "permanent fund dividends" into law to distribute a proportion of its oil earnings to Alaska residents. By the mid-1980s, with the collapse of oil prices, Alaskans were in a serious fiscal bind. In fact, they were saddled with an enormous debt, which Arlon Tussing described in the following terms:

> Alaska's oil revenues moved into the billions of dollars suddenly and unexpectedly in 1979, and by mid-1981 they were already in decline. Today, nobody knows where the bottom is or when it will be reached. Like the fortune of the proverbial sweepstakes winner or ex-champion, Alaska's oil wealth could disappear almost as quickly as it appeared, leaving the state broke and broken within a decade.[3]

The moral of that story is that you have to spread the revenues out when they come. A heritage fund, as was recommended for the Yukon by the Lysyk Commission, and as you've had in Alberta here for a long time, is one means of spreading the revenues out. We have to think of the future, because the good times will not last forever.

That no major development should occur before the interests of native northerners have been dealt with is now a generally accepted principle. Though far from perfectly, this principle has been addressed in the North American Arctic via land claims settlements and other arrangements.

A final point has to be made about the environment. Yes, indeed, we do have to be very careful about polluting Arctic waters and polluting Arctic lands by drilling for oil. Though there is legislation in place, this is a major problem, and I don't want to underemphasize it. The other thing we are finding is that we in the South, by virtue of driving our cars, by virtue of switching on our lights, by virtue of a whole lot of things, are sending pollutants north. The polychlorinated biphenyls (PCBs) found in mothers' milk in Broughton Island did not originate there. They weren't dumped there by an oil company. The methyl mercury flowing into Hudson Bay, and into who knows where else, comes from massive earthworks put in place to develop hydroelectricity here and to export that electricity from here to the United States. There is an effect, a spiral effect of some kind that I don't understand very well—I'm not a scientist—pushing airborne pollutants into the North: the Arctic haze. That is a far greater concern at this point than the odd oil spill, although I'm not trying to downplay the significance of oil spills.

It would be a tragic though plausible scenario if, after all the smoke of the debate had cleared, and northerners had indeed become masters in their own house with plenty of resource revenues flowing in, there was little of a useable environment left to them, not because of what had happened in their own backyards, but because of what had happened in the backyards down here in the South.

Notes

1. In August 1988, Dr. Colin Irwin submitted this report to Health and Welfare Canada, attracting considerable debate and media coverage. He drew attention to problems of high population growth, high unemployment, poor education and poor economic prospects for Inuit across the North. The underlying causes he identified included the failure of government programs to meet serious new challenges. As a result, controversy greeted Irwin's findings. However, he concluded his study by expressing confidence in the ability of the Inuit to solve their serious social problems if they are given more encouragement by government. See Dr. Colin Irwin, *Lords of the Arctic: Wards of the State. The Growing Inuit Population, Arctic Resettlement and their Effects on Social and Economic Change.* (Halifax, N.S.: Dalhousie University, 1988). (Editor)

2. The Northern Accords are agreements in principle between the federal and territorial governments, entered into in 1988, which allow the territorial governments to take over from the federal government the management of the oil and gas resource sector onshore, and the development of joint management regimes offshore north of 60° N. (Editor)

3. Arlon R. Tussing, "Alaska's Petroleum-Based Economy," in *Alaska Resources Development, Issues of the 1980s,* edited by Thomas A. Morehouse (Boulder, Co.: Westview Press, 1984).

Cooperation to Protect the Environment: The Finnish Initiative

Ambassador Esko Rajakoski

CHAPTER 6
Multilateral Cooperation to Protect the Arctic Environment: The Finnish Initiative

AMBASSADOR ESKO RAJAKOSKI

Stephen Lewis
We now turn to Ambassador Rajakoski of Finland, who is involved in the multilateral effort of his government to bring together numbers of countries in an initiative to protect the Arctic environment. I must say that this absolutely rings true. The one thing I learned more vividly than anything else at the United Nations is how extraordinary the Nordics are in their multilateral activities. If I was ever banned from this country for subversion, I have no doubt whatsoever that I would want to rest in one of the Nordic countries because they are exemplary in the way in which they care about this world. Ambassador Rajakoski.

Ambassador Rajakoski
May I first say how very pleased I am to have been invited to this conference and to have the opportunity to speak to this very distinguished audience. I am really amazed to see how large the audience is. That, I think, is strong testimony to the importance this country attaches to the Arctic.

At the outset, I would like to express my appreciation for the support Canada has given to the Finnish initiative, an initiative I am going to de-

Esko Rajakoski is Finland's Consultative Ambassador for Environment, Arctic and Antarctic Affairs. A master of political science of the University of Helsinki, he has held a variety of diplomatic and ambassadorial appointments in Moscow, Paris, Lima, London, Geneva and Buenos Aires, as well as at the United Nations. He holds the distinction of Commander, Order Lion of Finland, as well as 1st Class Order of Leopold, Belgium, and Commander of National Merit of France.

scribe later in my talk. When I speak of Canada, I have in mind the federal government, the scientific world, Inuit organizations, and many individuals, newspapers and other levels. I only now have come to know that the Finnish initiative itself is not new on this earth, having been started already by Mr. Berger some twelve or thirteen years ago. I am sure that the Finnish government would be very pleased to hire him to promote this idea.

Allow me first to say some words about the background of the concerns of the Finnish government and about the possibilities we see for future cooperation among the Arctic countries as far as the environment is concerned. As we have heard already today, and as we will see further this afternoon and tomorrow, there are several different factors that influence cooperation in the Arctic in general and environmental cooperation in particular.

The first factor is the military situation in the region, or, more generally, the factor of security policy. This factor, either directly or indirectly, has negatively affected the possibilities of cooperating in the field of the environment. This has been true through almost the whole period since the Second World War. The détente between the superpowers and the consequent lessening of tension in East-West relations have only lately provided more opportunities to enhance cooperation in the field of the environment. Security policy has been a negative factor, or at least a limiting factor, for cooperation in the field of environmental protection in the Arctic and is still a factor today.

The second factor is international law. An analysis of the situation in the Arctic from the point of view of international law, particularly one that keeps in mind the conventions, accords, treaties and other instruments entered into, shows that there is a clear need for cooperation in this field. Thus the element of international law is an exhortative factor towards the enhancement of cooperation.

The third factor is the exploitation of available natural resources in the Arctic, which has already taken place and which, sooner or later, is increasingly unavoidable. Exploitation should be carried out only after strictly taking into account the requirements of the protection of the environment. Economic reality is thus a forcing factor towards cooperation in the field of the environment.

The fourth factor is scientific cooperation. Scientific cooperation in the Arctic, both bilateral and multilateral, already has long traditions and should be followed up and further increased. We in Finland are encouraged to see the emergence of an international Arctic science committee, which will enhance the potential for multilateral efforts to increase scientific knowledge of Arctic nature. Thus the scientific element forms a supporting factor to cooperation in the field of the environment.

The strongest factor leading to the development of a new awareness of the worsening situation of the Arctic environment should be the pollution itself. There the facts speak their incorruptible language.

All these factors have something in common: each shows very clearly the lack of adequate multilateral international cooperation. This lack has been due mainly to the military situation in the Arctic Ocean, the tension between East and West, the lack of understanding and confidence among nations, and the fact that there are many international differences as far as the territories in this region are concerned.

But there is, fortunately, a basic change that has started to take place at an increasing speed. There are numerous examples of increasing interest in peaceful mutual cooperation. Important conferences and colloquia, such as this inquiry, have been arranged during the last two years or so. Scientific cooperation is increasing. Public opinion has been focusing on the Arctic more than ever. The mass media have addressed themselves to the Arctic and mainly to the problem of the protection of the environment. Actually, much has happened in a relatively short time. Mary Collins this morning mentioned a list of important happenings. I would also like to give you some of my own examples.

My first example is the creation of an international Arctic science committee. In 1986 at San Diego, during the meeting of the Scientific Committee on Antarctic Research (SCAR), some scientists invented the idea of creating an international organization for Arctic research activities. Within a couple of years, they elaborated the idea among the eight Arctic countries: Canada, the United States, the Soviet Union and the five Nordic countries—Finland, Sweden, Norway, Denmark and Iceland. Lately they have come to an agreement about the modalities of such an organization. It will, I hope, start its work very soon. I am sure that this international Arctic science committee is going to play an important role in the endeavours to preserve the Arctic environment.

My second example is the speech of General Secretary Gorbachev, which he made on October 1, 1987. In it, he presented six concrete proposals. As well, the speech indicated a new opening of the Soviet Union towards a cooperation that had often been lacking Soviet participation.

Another example of the chain of important gatherings that I should mention is the International Conference on Arctic Cooperation held in Toronto in October 1988, where substantial ground was covered as far as the Arctic is concerned. Another example is the huge conference in Leningrad, held from December 12 to 15, 1988, where almost five hundred scientists gathered to exchange notes about the results of their research work.

Mary Simon was talking some minutes ago about the activities of the Inuit Circumpolar Conference. We think that it has a very important role in these endeavours. I should also mention that parliamentarians of seven countries assembled in Moscow last month [February 1989] to discuss the same Arctic questions. And, as we have seen, the press has been increasingly interested in cooperation in the Arctic. I must say that this inquiry fits in a very timely way into this general picture.

The Finnish government analyzed the factors I have described and took into account the fact that the equally fragile Antarctic environment has been taken care of fairly effectively through the Antarctic Treaty system. It came to the conclusion that it is high time the eight Arctic countries take coordinated multilateral action to tackle the problem of the Arctic environment.

All this is very recent. It was only last October [1988] that representatives from our government started consultations with the other seven Arctic countries in order to identify the possibilities for common action. The results have all been positive; they have been very encouraging indeed.

On the basis of these consultations, the Finnish government took an official decision to launch this proposal on January 12 of this year [1989], that is to say, only a couple of months ago. The proposal was sent to the ministers for foreign affairs of the seven other Arctic countries.

The proposal was signed by two ministers of the Finnish government, the Minister of Foreign Affairs and the Minister of the Environment. There was of course a message in this not very common formula. The first part of the message is that it is the Minister of Foreign Affairs who is going to take care of the negotiations, because we are fully aware of all the political difficulties, the political connotations, this proposal carries with it. At the same time, we wanted to underline that this is an environmental issue, which we did by using the signature of the Minister of the Environment. We do not want to touch those very delicate international problems that have other than an environmental character.

The second part of the Finnish government's proposal is a working paper in which we discuss both the substance of the proposal and certain details. The working paper starts from the fact that the ecosystem of the Arctic is very fragile. Due to the extreme climatic and ecological conditions, the flora and the microorganisms in this area can only very slowly be renewed or, after disturbances of the Arctic equilibrium, revived. The adverse impact of human activities has, in addition, considerably increased in the Arctic, especially during this decade, so that the area has become exposed to danger. So far, no comprehensive multilateral protection measures have been undertaken to safeguard the natural heritage of Arctic resources. It is for those reasons that the government of Finland thought that urgent measures to protect the Arctic environment are necessary at the earliest possible time.

The working paper identifies the most important threats to the Arctic environment today. These are climate change, pollution of the marine environment, and exploitation of the living and nonrenewable resources. These threats have a direct impact on the well-being, traditions and cultures of the Arctic peoples. The consequences of the climatic change to the Arctic environment also need to be acknowledged. The emissions of certain air-polluting compounds released from the territories of the states bordering the Arctic evidently have adverse effects on this region. In addition, long-range, transboundary air pollution poses an increasing risk both to the land and sea areas in the Arctic.

In the working paper it is pointed out that the ecological equilibrium, the resources and the legitimate uses of the marine and coastal environment are threatened by pollution and by insufficient integration of environmental concerns into the development process. Effective measures are needed to enhance the protection of the Arctic sea areas and to protect them both from pollution caused by shipping, oil and gas drilling, and mining, and from pollution from land-based sources including rivers, estuaries, outfalls and pipelines. Attention should also be paid to pollution caused by dumping wastes and other matters into the sea, and to the protection of polynyas[1] where both human activities and wildlife are concentrated. A particular problem is caused by pollutants transported along sea currents from the Atlantic and the Pacific Oceans.

The value of the living resources of the Arctic needs to be acknowledged. It is therefore important to seek to ensure that resource development is in harmony with both the maintenance of the unique environmental quality of the region and the evolving principles of sustained resource management. The development of the living resources in the Arctic should be based on the principles of the World Conservation Strategy.[2]

It is further stated in the working paper that it is important to include living resources in the protection program: marine, coastal and archeologic ecosystems, and ecosystems in estuaries, mountains, tundra and northern boreal forests, including peat bogs. The whole Arctic region would also be affected by radioactive contamination caused by emissions from nuclear power plants, nuclear field vessels or processing plants, or by nuclear accidents. Therefore, effective protection of the Arctic requires development of three things: intergovernmental cooperation, scientific research and monitoring of the ecosystems. It is important to further develop and implement a scientific program, as well as to cooperate with the relevant existing scientific organizations. The Arctic states should facilitate the conduct of scientific activities in the Arctic and coordinate their scientific work.

The Finnish proposal includes an invitation to take specific intergovernmental measures for the establishment of criteria and standards concerning human activities that have an impact on the Arctic environment. Furthermore, effective notification and consultation systems should be developed for dealing with situations in which the marine environment is in imminent danger of being damaged by pollution. We also laid some groundwork in international law and came to the conclusion that regimes for the regulation of human activities having an impact on the Arctic environment fall into four groups.

First, there is a group of global conventions, which, because of lack of time, I will not describe. Second, there is a group of regional conventions that regulate activities that have an impact on the Arctic. These include the convention for the prevention of marine pollution by dumping from ships and aircraft, the so-called Oslo Convention. Third, I would mention the Paris Convention and the Helsinki Convention on the protection of the marine environment in the Baltic Sea area. Fourth is a group of agreements aimed at protecting Arctic wildlife, namely whales (1946), polar bears (1973) and North Atlantic seals (1957 and 1971).

In conclusion—and I think this is important—it can be observed that there exists no comprehensive regime concerning the conduct of human activities having an adverse impact on the Arctic environment or its resources. For the reasons I have given, the government of Finland deems it necessary to initiate an intergovernmental process with a view to elaborating coordinated and concerted action for the protection of the Arctic environment. This process could lead, for example, to a declaration, convention or other multilateral arrangement exclusively dedicated to the protection of the environment of the Arctic. We have also proposed that such action should be taken by the eight countries that possess sea and land areas north of the Arctic Circle. Such action should not contradict existing multilateral or bilateral agreements covering human activities affecting this area.

We think it is extremely important to tackle the problem of our very fragile environment and to start doing something to protect it. I have not gone into many interesting scientific details, which should be mentioned for a better understanding of the urgency of the matter. I hope our scientific speakers will do so. I will only say this: I do not think it is too late to tackle the question of the Arctic environment. I am convinced that humankind is able to use its abilities to develop new technologies, which will serve the environment as well as they have been able to destroy and pollute it. The question is: Are we ready to recognize that much of our future, if not the whole existence of mankind, depends on our behaviour vis-à-vis nature? The question is: Are we ready to take action mutually to protect the environment? The question is: In this particular situation, are we ready to concert and coordinate our efforts to save the Arctic environment? I hope we are. That is the hope of the Finnish government.

Notes

1. Polynyas are persistent or recurrent areas of open water in fast ice or ice-choked seas. (Editor)

2. The World Conservation Strategy is a generalized statement of conservation goals and the means to achieve them, prepared as a collaborative project by the International Union for Conservation of Nature and Natural Resources (IUCN) with advice, cooperation and funding from the United Nations Environmental Program and the World Wildlife Fund. See *World Conservation Strategy. Living Resource Conservation for Sustainable Development*, IUCN, (Switzerland: 1980). (Editor)

Question Session

Stephen Lewis

CHAPTER 7
Question Session

THOMAS R. BERGER, JOHN F. MERRITT,
AMBASSADOR ESKO RAJAKOSKI,
MARY SIMON AND EDWARD R. WEICK

Members of the media panel and the audience had an opportunity to question the speakers. The media panel consisted of Gwynne Dyer, Linda Hughes and Ann Medina. Stephen Lewis was the moderator on the first day of the inquiry; Adrienne Clarkson was the moderator on the second day.

Linda Hughes
Mary Simon, you said that the Inuit feel strongly that they have been excluded from the policy-making process and that they want to reverse the

Stephen Lewis holds the Barker Fairley Distinguished Visitorship in Canadian Culture at University College, University of Toronto. He is also special advisor on Africa to the Secretary General of the United Nations. He was educated at the universities of Toronto and British Columbia, after which he spent a year teaching and travelling in Africa. He was elected to the legislative assembly of Ontario in 1963 at the age of 25, and after four successive re-elections was elected provincial leader of the New Democratic Party. He stepped down after two years and resigned his seat in 1978, embarking on a career as a radio and television commentator, and labour relations arbitrator. Mr. Lewis was named Canada's ambassador to the United Nations in 1984. He resigned this position in 1988. He holds a dozen honorary degrees, the Gordon Sinclair ACTRA Award for "outspoken opinion and integrity" in broadcasting, and other awards including the Human Relations Award from the Canadian Council of Christians and Jews "for his outstanding contribution to understanding and respect among Canadians."

Question Session

Gwynne Dyer

trend towards militarization. How do you think the Inuit can be brought into the policy-making and decision-making process in this country?

Mary Simon
There are a number of ways. Through our nongovernmental organizations, such as the Inuit Circumpolar Conference (ICC), we are attempting to work with the Canadian government on policy-making; but the problem we have in the Arctic is that we don't get very much information on what is happening, so we cannot make any informed analysis or decisions on Arctic militarization. In Greenland we have a relationship with the home rule government, which works fairly well, although it is not the home rule government that is responsible for defence and foreign affairs, but Denmark. Because the home rule government does have a very active role in policy development, the Inuit in Greenland do have some input, but they have none at all in Canada or in Alaska.

Hughes
Would provincial status for the Northwest Territories make a difference?

Simon
Yes, it would.

Gwynne Dyer
May I ask Mary Simon a question also? This has to do with the scope of the Inuit Circumpolar Conference. Has there been any progress whatsoever on including the Inuit of the eastern Soviet Arctic in the ICC, in giving them access to their own government and to our government as far as your organization is concerned? Secondly, you spoke about the home rule government in Greenland; it seems to be the one group of Inuit that has some possibility of moving ahead on demilitarization on its own. Has there been any discussion of that?

Simon
Inuit from the Soviet Union will be participating in the ICC general as-

Gwynne Dyer is a syndicated columnist specializing in international affairs. His twice-weekly columns are published by 150 newspapers in 30 countries. He has served as a reserve officer in the navies of Canada, the United States and Britain. He holds degrees in history from Memorial and Rice Universities, and a Ph.D. from the University of London. He has lectured on military history and war studies. His works include two books, War *(New York: Crown, and London: Bodley Head, 1985) and* The Defence of Canada *(forthcoming), and three films.*

Question Session

Linda Hughes

sembly taking place this July. I was over there in the summer of 1988. We negotiated an agreement with the Soviet authorities and there will be approximately twenty-five to thirty people attending the general assembly. In addition to that, I've had some discussions with the Soviet authorities on the idea of a zone of peace, because Mr. Gorbachev talked about it in his speech in Murmansk in 1987, and they are very open to receiving proposals from us. We've got a project that is called a nuclear-weapons-free-zone project. We've had funding problems, so that project hasn't gone very far.

In regard to your second question, the home rule government works quite closely with Denmark, because Denmark is responsible for Greenland's defence and foreign affairs. The home rule government has just created a committee of ministers to address the issue of security in the Arctic, and I understand that they've had several high level meetings. One of the trips they are taking is to Colorado to visit the North American Aerospace Command (NORAD). They are becoming much more involved.

Ann Medina
I must say I'm a bit disappointed. It was my understanding that we would have all the speakers here; and I think there would have been some fascinating discussion, Mary, between you and Mary Collins in terms of some of the comments you have made. Is it still at all possible to bring her back, Stephen? I hate to put you on the spot here.

Stephen Lewis
You are not putting me on the spot. My understanding is that indeed the questions were to be confined to John Merritt and Mary Simon, and that the Minister was not a part of the panel responding to questions. That was arranged with the organizers of the conference. So though you might wish it otherwise, it is your organizers who arranged that; so you will have to ask questions of John and Mary.

Linda Hughes is editor of The Edmonton Journal. *After receiving an honours degree in history and economics from the University of Victoria, she joined the* Victoria Daily Times, *where she became an award-winning reporter covering education and provincial politics. She joined* The Edmonton Journal *in 1976, covering civic politics. In 1977 she won a Southam Fellowship and studied political science and economics at the University of Toronto for a year. She joined* The Edmonton Journal's *editorial board in 1978 and served in several progressively more senior positions before being appointed editor-in-chief in March 1987.*

Question Session

Ann Medina

Medina

Mary Simon, one of the points that has been discussed is whether the approach to the European disarmament question is separate or part of the whole question of the Arctic. What specifically seems to be the difference between your position and that of the Canadian government? Is this a word war?

Simon

The ICC believes that you can't deal with the two in isolation but that they go hand in hand. Security in the Arctic is tied into all the activities being undertaken right now, so I think that it's linked.

Medina

The government, as I understand it, would say that the two are linked but also separate. So what specifically seems to be the difference? It sounded as if Joe Clark was making a bit of a stronger statement when you were quoting him—that there is a question of isolation—but you were saying you don't feel that it should be isolated.

Simon

We feel that Arctic security includes environmental, economic and cultural, as well as defence, aspects. We feel that a more reasonable position would have been to recognize fully all the aspects of Arctic security, while adding that the North Atlantic Treaty Organization (NATO) might be consulted on the military aspects.

Medina

I would like to go into a question following up on the Soviet Union. Al-

Ann Medina is a journalist and producer, currently making her first motion picture, and Senior Resident of the newly inaugurated Canadian Centre for Advanced Film Studies. She has a B.A. from Wellesley College and an M.A. from the University of Chicago, in philosophy. She has also studied at Harvard University and the University of Edinburgh. She began her television career with NBC in Chicago, later working in Cleveland and New York. She moved to Canada in 1974 and joined the CBC's Newsmagazine, *first as a reporter and later as executive producer. During this period and subsequently as senior journalist and producer with the CBC's* The Journal, *she travelled and reported extensively from abroad, mainly from the Middle East; she served for a period as the CBC's Beirut bureau chief. She has been awarded an Emmy for "outstanding individual achievement," as well as several film festival awards.*

though you were talking about the Inuit who are in Soviet territory being included, what are the differences between the kinds of reactions that you get from the Soviet Union and, say, Canada?

Simon
The difference in reaction I got when I was in the Soviet Union in August was that the authorities were very open to discussing our ideas on Arctic security with us. They want to develop more of an open dialogue, especially in the area of creating a zone of peace. They know that we are very much concerned about that idea and that the ICC passed a resolution in 1986 calling on the ICC executive council to work on a project to try to come up with a proposal that could be presented to governments. In terms of the Inuit involvement in Siberia, I think this is just the beginning. We are, I think, much more ahead in dealing with these issues and confronting government on our concerns and our views; this is just beginning in the Soviet Union.

Hughes
I too wish that Mary Collins were here, because this question would probably be answered by her, but I'll ask you instead, since you're here, Mary Simon. What has the Canadian government's response been to Mr. Gorbachev's Murmansk initiative, which was I believe in 1987, and which includes the issue of the Arctic as a zone of peace? Has there been an official response from the federal government, or has your organization discussed it with the federal government?

Simon
We brought it to the attention of the government, but I don't know what the position of the federal government is on it.

Hughes
Did they respond to the Murmansk initiative?

Simon
I don't think so.

John F. Merritt
I'd like to add something about the response to the Murmansk speech. As far as I know the only responses have been fairly dismissive remarks by a couple of ministers, which I frankly find very disappointing. Not that Canada should accept at face value what the foreign minister or leader of any other state says on any topic, but just to react by saying that it can't be taken seriously in the absence of the Soviets' demilitarizing unilaterally in

the Kola Peninsula is, I think, not taking seriously what potentially could be a very serious proposal. I think, in a broader sense, one of the consequences of the Intermediate-range Nuclear Forces (INF) agreement will be that there is greater and greater superpower interest in naval arms buildup in the Arctic. There are technological reasons for that as well, relating to cruise missiles and submarines. In one conversation I had with an American arms control expert not long ago, he suggested that even today the two superpowers could deploy as many as seventeen thousand naval nuclear warheads in Arctic waters. So the INF agreement is great, but if it means that the arms race is being exported to the Arctic, that has major implications for Canada; and I would certainly support Mary Simon's point of view that it is not adequate for Canada to defer all these questions to Moscow and Washington.

Dyer
Could I ask John Merritt a question? I know that things have been moving very fast recently on developments in the ozone layer investigation in the Arctic, as differentiated from the Antarctic, and that a hole in the ozone was allegedly discovered over Baffin Island for the first time last month. Can you give me a worst-case estimate, or guesstimate, of how urgent the depletion of the ozone layer is, in terms of the health of people living in the Arctic? If you can, though I realize this would be an even bolder guesstimate, would you give an estimate of how soon really drastic climatic changes could occur in the Arctic as a result of carbon dioxide levels, greenhouse effect and so on?

Merritt
To answer the second question first, there is obviously a lot of arguing going on about the various models used to predict climatic change, but my impression from talking to Department of Environment scientists, even more than a year or two ago, is that they are now fairly confident that these things are going to show up early in the next century. We are talking about the possibility of major global temperature change by, say, the year 2025 or 2050; and that's not very far away. The frightening thing for the Arctic, as I've mentioned in my presentation, is that fairly modest changes in worldwide average temperatures will show up in a very dramatic way in the high latitudes. There are various reasons for that. I'm not entirely clear what they are—I don't think the scientists are—but unfortunately the Arctic has particular properties relating to ocean current patterns and windflow patterns that make it, in some ways, the final repository for a lot of the world's environmental problems, a lot of the world's pollution. The scary stories now coming out of Baffin Island relating to toxins in the milk of nursing mothers have a rather ironic side

Question Session

to them. A Laval University study team went there to try to find a control population, which they were confident would have virtually no traces of these toxins, only to find rates that appear to be higher than anywhere else in the world. So the Arctic does act as a sink and unfortunately it means that the people who live there may well inherit all the environmental problems generated elsewhere.

To go to your first question, the worst-case scenario for the ozone layer, I'm not particularly well versed in the science associated with that. I suppose the worst-case scenario would be that it will be dangerous for Inuit to go outside for extended periods of time; and I don't say that facetiously, because a lot of people spend a lot of time outdoors if they are living a traditional lifestyle. And related to that, the worst-case scenario would be that it is no longer possible to live on what is called "country food," so that people would become entirely dependent on imported foodstuffs. And of course you can imagine all the social problems that come with people being forced to be sedentary in communities and living off imported products when many of them in the past have been self-supporting, self-reliant and living large parts of their lives outdoors away from the stresses of community life. So the worst-case scenarios are very bad. I certainly couldn't attach any kind of timetable or give you a risk factor there.

Medina
This is just a brief one. Could you fill us in on the current status of the Grand Canal project,[1] and perhaps, Mary Simon, could you give us your reaction to it?

Merritt
I don't think you are going to see a spade in the ground in the foreseeable future, but like a lot of these proposals, it is there and it kind of percolates along, and my impression from the last couple of years of my work is that a lot of the development scenarios, which at some point seem to be exploded, have a way of coming back again.

An example of this is the pipeline proposed for the North Slope of the Yukon. Ten years ago it looked pretty certain that Canada was fully committed against the prospect of a North Slope Yukon pipeline through some very environmentally sensitive areas in an area that is now a national park. A year and a half ago we had a parliamentary committee recommending that route once again, and although federal ministers at that time were fairly quick to assure organizations like ours that these proposals had no standing, the recent activity over the last few weeks on the proposed Mackenzie Valley pipeline makes me fairly confident in saying that the battle for the North Slope of the Yukon will have to be fought again.

So I think these schemes should never be viewed as far out and impossible and never going to happen.

Lewis
Let me, if I may, say two things. One—just as a footnote to what Gwynne Dyer raised—I've attended two fascinating conferences in the last number of months on climate change. The cataclysmic predictions, if, as John says, the change is effected early in the twenty-first century, are really quite unnerving. If they are cataclysmic for the world generally, they are positively apocalyptic for the Arctic. And it is worth noting that Gro Harlem Brundtland herself has made analogies to consequences of climate change and consequences of nuclear war without any of the many scientists in the audience expressing a dissenting view. So we are, as you recognize, truly dealing with pressing problems.

Secondly, it has been drawn to my attention that tomorrow Ambassador Roche and Major-General Huddleston can respond on behalf of the government to specific questions that arise during the course of the inquiry.

Medina
Mr. Berger, how do you feel about Mr. Weick's assessment of the urgency of economic development in the northern communities? Second, given your notion of what development means in terms of strengthening traditional economies, how do you think that kind of development could be paid for, since money was identified as the key in Mr. Weick's presentation?

Thomas R. Berger
I don't think that Ed Weick and I are really very far apart on this. I've never thought of it as an either/or proposition.

Alberta has its oil and gas industry. It nevertheless has what you might call a subsistence economy, that is, agriculture, which persists on a sustainable basis and will persist through the years to come when there is no more oil and gas or the price of hydrocarbons hits rock bottom. I have always thought of the subsistence economy—the traditional hunting, trapping and fishing economy—of the North and the rural and frontier areas of Canada in that way. That is what native people depended on before oil and gas came and that is what they will have to depend on when oil and gas are gone. And that is why I've always made a nuisance of myself by harping on the subject.

I went to Alaska for two years, from 1983 to 1985, to conduct a land claims review, and I visited sixty villages in the bush, Indian and Inuit villages. I found there, notwithstanding the great oil development of the

1970s and early 1980s and the enormous flood of cash it brought to Alaska, that the subsistence economy was more important than ever and thriving to a greater extent than anyone would have thought possible. That is because the farther north you go, the more limited conventional economic opportunities become. Geography, climate and distance are all against entrepreneurial success as we know it.

Ed suggested, and he is perfectly correct, that if you engage in subsistence activities—hunting, fishing and trapping—you need cash. You need to pay for your snowmobile, you need to pay for food for your dogs, or you need cash to charter an aircraft to get out to the hunting grounds. Northern people have obtained that cash often through working part time (that is, for three or four months of the year on the oil rigs or perhaps working for the government). It is a kind of lifestyle that we are not used to because we think of people being obliged to carry a briefcase or a lunchpail and that is what you do year-round. You don't chop and change. But as far as obtaining the cash for this kind of activity is concerned, it has been built into the James Bay and Northern Quebec Agreement. The Crees arranged that part of their land claims money would go into a fund to provide a guaranteed annual income to Cree families so that they could continue to live and hunt in the bush at certain times of year, and from every report, that has been the greatest success of the James Bay and Northern Quebec Agreement.

I understand that the Dene and the Métis are trying to build the very same provision into their land claims agreement in the Mackenzie Valley. So I don't think these things are incompatible.

I want people to understand that when you talk about oil and gas development and pipelines you should not get stars in your eyes about the number of jobs it will bring northerners. There will be economic spinoffs; there will undoubtedly be cash benefits; but at the end of the day, as in Alaska, you might find very few native people actually employed in the industry.

Medina
I was a bit confused by Mr. Weick's arithmetic. It seemed initially you were making the argument that resource development will pay for much of the development the northern communities need, but then you said that, in the end, the communities were fairly substantially in debt. So it seems that fiscal arguments work against that kind of resource development.

Edward R. Weick
No. I mentioned the debt problem in Alaska and in the North Slope Borough. What happened was that the pace of local development outran the

revenue available, which also happens to many municipalities. In fact it has happened to our national government. So you acquire a big debt. There is no great mystery about that.

What I suggested as a remedy is that, sooner or later, you learn that you do have to put things away for a rainy day and that the oil and gas industry in any particular region is there for a while and is not necessarily there forever. I was also distinguishing between two kinds of financial needs. One is a major infrastructure need for schools, hospitals and all the things that higher levels of governments have to provide. In the territories all these things over the past several decades have been paid for very largely through federal transfers because there is no substantial revenue base in the North. The base is in fact quite small. So that is one kind of need for cash.

The other kind of need, that Tom Berger was just referring to, is for cash to carry on your economy. In the North, that is a subsistence economy. In the South, in some regions, it is a farming economy, and many farmers have had to go and work in industrial jobs to keep their farms going. But in the North there is a continuous need for cash inputs into the subsistence economy. This economy has changed a great deal since the days of the dog team. The only ways you can get that cash now, it seems, with the decline of the sealing economy and the fur economy—though they may come back—is through wage work or through government grants or through welfare. I think that the least demeaning way of getting it is through wage work. So there is that sort of trade-off, that coexistence, between wage employment and the subsistence economy now, which has become quite built-in.

Hughes
In a recent news story that *The Edmonton Journal* carried about the National Energy Board hearings on Mackenzie Delta gas reserves, we quoted an oil company as saying that it is confident that native opposition has lessened or been eliminated this time around. In fact I think the quote was that there had been a "maturing process" among the people involved, and I think the implications of that are probably pretty offensive. But what I am wondering, Mr. Berger, is whether you think that this time around there will be a major change in the response of the indigenous peoples and whether their concerns will perhaps now focus more on the environment than on economic issues.

Berger
Well, these hearings remain to be played out. I don't speak for northern peoples, but when I held those hearings back in the late 1970s, we heard from about one thousand native witnesses, ninety-nine percent of whom

opposed the pipeline because they thought that the social and economic impact would be devastating; they thought it would put an end to the possibility of settling their land claims on favourable terms; and they were concerned about the impact on the subsistence economy.

Now they have settled their land claims. The Dene and Métis settlement is only a settlement in principle, and the land selections haven't been made, but the spokesmen for these organizations indicate that they feel it can be completed within two or three years. The subsistence economy has been strengthened. The provisions being made to secure access to fish and wildlife resources exclusively for native people in the North are the most significant provisions of the Dene-Métis agreement, as I read it. And so I think there will not be the same united front opposed to the pipeline that there was last time. Bear in mind that native people now control the territorial assembly, and Ottawa has devolved many federal powers to the territorial government, which could not exercise them in the 1970s.

So I think it is a new ball game. And from an environmental point of view, as I said earlier, there is a second volume to my 1977 report that shows exactly how you can build a pipeline in the Mackenzie Valley with appropriate environmental safeguards. I don't think that there is any overriding environmental reason not to build it. I think the real question may well be what we have in the way of gas reserves and whether we can afford to export this gas. Those are national questions, not purely northern questions.

Dyer
May I ask Ambassador Rajakoski a question? I gathered from the way you described having avoided raising other than environmental questions in your proposal for cooperation among the eight Arctic nations that you felt there was not much prospect of embracing security considerations within this proposal as well. Would you confirm the hypothesis that the perception was that there is no point in trying to deal with security considerations under the same umbrella? If that is the case, what prospects do you see for some parallel proposal that would address security considerations in the Arctic among the eight Arctic nations?

Esko Rajakoski
We all know how very difficult it has been to set up any multilateral cooperation in the field of the Arctic. So I think it's prudent to start from that very urgent point, the protection of the environment. We think, from our analysis, that putting in any other elements that would have a political connotation or would deal with the differences—disputes actually—that exist in the area could only hamper this very important initiative. What

we are aiming at here, to state it very briefly, is to start a kind of cooperation in which the eight Arctic governments could make a political commitment to do something about the Arctic. To have the right regime and have the right decisions locally, it is basic that you need to know what is going on; you need to have the scientific research data. Only after that can you deal with these problems appropriately. The third thing that is very important here is the monitoring. It is not only the exchange of information, but the monitoring of the situation in order to see where we are going, which decisions should be revised, or what new actions should be taken.

Berger
I would like to add a postscript to what Ambassador Rajakoski said. The Ambassador mentioned the polar bear treaty of 1973. That is a good example of what can be achieved. There was a concern about the decline in polar bear populations throughout the circumpolar basin in the Soviet Union, the Nordic countries, Alaska, Canada and Greenland.[2] In 1973 a treaty was made among all those powers, the Soviet Union, the United States, Canada and the others, and it has been working now for sixteen years. It has meant that polar bear populations have been preserved and enhanced. It shows that the kind of larger proposal Finland has in mind can be achieved if the will is there.

Jerry Paschen (Canadians for Responsible Northern Development, Edmonton, Alberta)
The question is to Ed Weick. He elaborated in detail on the deficit, the royalties on petroleum and natural gas resource development in the North, as well as pollution and the environment. What he forgot to say, and I would like to say, is that the present application before the National Energy Board is basically a gas export application to the United States. What he also forgot to say is that we have to look at the Canadian North not only as a place to extract resources, such as petroleum, but we have to look at the issue of responsible northern development for a sustainable future.

My question is: In view of all these facts, and in view of the fact that no mention was made in the gas export application about an economic northern development plan, I would like to find out what Ed Weick's plan is for the Canadian North in view of the fact that it is a large storehouse of minerals; that is, zinc, lead, iron and numerous other ones.

Weick
The North is not a storehouse. What you can foresee is eventual exploitation of its resources—at least those that are economic—but probably not

all of them. I don't believe I neglected to say that the gas export application involves gas going to the United States; I said something about a continental energy market, which I believe, whether we like it or not, we are into.

I think the sustainability question has to be addressed through what you do with the revenues. First, you have to get a cut of the action. That's probably the most important thing. Then, once you have established that, by means of Northern Accords or by means of land claims agreements or whatever, you have to decide what you are going to do with the money. If you spend it all at once or spend it foolishly, then you haven't got it. Sustainability is the result of a planning exercise that has to involve long-term thinking.

The one thing that I am quite convinced of is that, in order to build all the things you need in a modern economy, you do need large sources of funds. I would like somebody to tell me where else, other than through nonrenewable resource exploitation, those funds can come from in the North. I mean very large volumes of funds, billions of dollars, because that is what it costs.

Laureen Pameolik (Qitiqliq School, Arviat, Northwest Territories)[3]
Arvianuvugut.[4] We are not speaking our native language to be rude but to make a point. Our language, our people and our land are one. If you don't understand us, a few trips to Tuktoyaktuk are not enough to help you understand the North. The Northwest Territories has become a welfare territory. We used to be self-sufficient, but now our community of thirteen hundred people has about seventy-five percent unemployment. We need southern tax dollars to feed ourselves.

Sam Alagalak (Qitiqliq School, Arviat, Northwest Territories)
The only kind of economic development the Canadian government seems willing to support is uranium mining, offshore drilling and military expansion. All of these are unacceptable to us Inuit. The Meech Lake Accord will guarantee our political dependence. We would like to ask Mary Simon how we Inuit can safeguard our land when we have no economic or political power.

Simon
Inuttitut tusariamitquvianapuk.[5] I'll answer in English for the benefit of the audience. The question you ask is related to what we are trying to achieve as Inuit, and it relates to the issue of being a self-determining people, the issue of self-government. We have been trying to achieve a certain autonomy for a number of years now, and we continue through the land claims process and through constitutional negotiations that have now

ended but which we would like to see continue. I think that if we can participate and be partners in Canadian society when it comes to not only economic development but also political development we will be able to conserve our lifestyles, our economy and our environment.

Pritam Atwal (Beaver Hill Peace Group)
It appears that the Arctic peoples have no choice but to develop, and their traditional ways of living, which are derogatorily termed "subsistence economies," are not seen as being adequate. Has there been any serious attempt to define what development is really desirable for the indigenous Arctic people? My question is addressed to Tom Berger please.

Berger
In Alaska, the Indian and Inuit people call their traditional economy the subsistence economy. "Subsistence" doesn't have the kind of derogatory implications that I think you derive from the term. It is the way they live.

Could I just say that the discussion here this morning is focused to perhaps an inordinate extent on the whole question of subsidizing the subsistence economy and where the money is going to come from. It is thought that it is somehow inappropriate that the money should come from government, that there must be megaprojects in the Arctic to supply the revenue to subsidize the subsistence economy. I don't know where that idea comes from. You know here in the Prairies you have primary producers, the farmers. And it is thought that that way of life is worth preserving, that it is important to the integrity of the communities where farm people live; they are subsidized by the rest of us and we don't complain. On the West Coast and the East Coast we subsidize primary producers—we subsidize people engaged in fishing so that, in the fishing villages, families may still live a way of life we think worth preserving and enhancing; they are subsidized through unemployment insurance. The idea that the native people in the Arctic and Subarctic engaged in subsistence hunting, fishing and trapping activity should be subsidized, it seems to me, is entirely consistent with the Canadian ethic. We don't believe in the American notion of social Darwinism. We subsidize each other. That is part of what this Canadian ideal is about.

John Plaice (University of Victoria, British Columbia)
If the entire world were to stop burning fossil fuels this very day, the proportion of carbon dioxide in the atmosphere would, by the year 2050, be double the proportion of 1900. The most optimistic models talk about a resulting temperature increase of one and a half degrees; the more pessimistic, more than four degrees. A two to four degree increase could well mean the demise of the caribou in the very near future, perhaps with-

in ten years. In the next twenty years we will be forced to look for alternative sources and currencies of energy. By doing oil and gas exploration we are spending billions of dollars, all to no avail. All this to saddle the northerners with more debt. I would say that there *is* an overriding environmental reason for no pipeline, the greenhouse effect. Justice Berger, how do you respond?

Berger
I am a private citizen, and I am making a living as a lawyer in Vancouver, and I try to follow all these things, but I don't pretend to be able to predict the implications of the greenhouse effect. In my talk this morning I stayed with immediate concerns of mine: the Porcupine caribou herd and the implications of offshore oil and gas exploration and development as they affect Arctic waters and marine life and weather systems. If Canadians as a whole decided that bringing gas from the Arctic for export to the United States was, in a global sense, environmentally unsound, well that is something that they must consider very seriously. But forgive me, I just don't know enough about it to offer a firm point of view.

Saul Arbess (Regina, Saskatchewan)
This question relates to the comments made by the Inuit speakers earlier and also, I think, to something that Thomas Berger just said. This is for Mary Simon. Do you feel that the question of the relationship between the circumpolar regions and the Eastern and Western state powers would be enhanced if we conceptualized the issues and pursued the dialogue in terms of a North-South framework rather than an East-West framework? If so, what mechanisms do you foresee in order that your voice will be heard?

Simon
We Inuit feel that it is very important that a dialogue be undertaken, not only between the Inuit around the circumpolar region—which includes the Soviet Inuit—but that dialogue be undertaken between the North and the South. As I was saying earlier, there are specific things that should be undertaken, one of which is that there has to be a clear response to our request to be involved and to participate in the policy-making process. I think that if the relationship in terms of North-South were developed as one of those premises, we would be going a long way towards promoting peace and security.

Berger
Could I add to what Mary said? The Inuit Circumpolar Conference (ICC) is an organization of Inuit people in the circumpolar region. They have, I

think, led the way for all of us, including the nation-states, in establishing a framework for circumpolar cooperation. I know a little bit about them because the land claims review I did in Alaska from 1983 to 1985 was for the ICC. Mary, I am sure, will speak with conviction about this. On a shoestring budget and without the funding that the nation-states have for so many projects, the ICC has made its presence felt throughout the Arctic and Subarctic. It has come out in favour of a nuclear-free Arctic. If they can do that—these people getting together from vast distances—surely the nation-states can be expected to perform a little more adequately than they have in the past.

Bob McCardle (New Sarepta, Alberta)
Mr. Weick, you seem to be suggesting that settling land claims and native self-government are incompatible with long-term environmental protection. We have also heard that the North is very quickly becoming a toxic waste dump. Don't you think that you seriously underestimate the knowledge and the ability of Inuit and other native groups to know and decide for themselves what types of development are desired and are environmentally sound for northern latitudes?

Weick
I'm not sure; what have you got me suggesting again?

McCardle
You seem to be suggesting that settling land claims and native self-government are incompatible with long-term environmental protection.

Weick
The world has shifted a bit. I think maybe ten or fifteen years ago there was a tendency to align native interests and environmental interests very strongly. Native interests, I think, have shifted over to the development side to some extent, but that does not necessarily mean that they are anti-environmental in any sense.

I think we have come to realize that the environment has to be protected however people align themselves or whatever political arrangements are made. The environment is a constant in the whole thing. And just because native people now have a stronger stake in resource development than they had fifteen years ago, it does not follow at all that they have a lesser stake in the environment.

Juliet Knapton (Edmonton, Alberta)
Mary Simon, you mention a lack of information for native peoples to make decisions. Does this apply to all the Arctic countries and areas?

Question Session

Also, if the people do have the information that Mr. Weick mentioned about industrial impact from society, does such a thing as a choice of having a VCR amount to demonstrating a choice of a way of life?

Simon
The problem of lack of information applies primarily to the Canadian Arctic and to Alaska—the Alaskan Inuit are confronted with the same problem. It is less of a problem in Greenland because they have their own home rule government, the government there is run by Inuit, and they are able to communicate much more freely in their own language with the public.

In northern Canada, organizations like the ICC are attempting to provide a way to communicate with northerners so that they are more informed on activities such as militarization and resource development. This is because there is such a lack of information that they are not able even to give an opinion on the situation that is confronting them. In order for the people to make more informed decisions, there has to be much more access to information in the North. So the problem applies to most of the Arctic. I wasn't sure about the second part of your question.

Knapton
Just that Mr. Weick had said that, for example, to have a VCR is to choose a way of life or to demonstrate something.

Lewis
Mr. Weick is shaking his head.

Weick
Did I say VCR?

Lewis
No, you didn't say VCR.

Weick
I'm glad I didn't. What I was saying was that, yes, indeed, by buying up a transportation company, by owning a drilling company and so on, people are demonstrating something. They are demonstrating that they have made a commitment of some kind. Now that does not mean at all, even remotely, that they have abandoned other commitments. I wasn't talking about VCRs. I think we all have them or would like to have them.

Lewis
The questioner, however, was putting a vivid construction on your suggestion, which was perhaps a logical extension of the suggestion.

Chris Labossiere (Holy Trinity Catholic High School, Edmonton, Alberta)
Mr. Weick, as you mentioned, northern resource development is inevitable and the economic boost it would bring to the northern regions is quite blatant. With your experience at Dome Petroleum and the Berger Commission, would you say that major corporations of today could develop an area of the North economically, and through this socially, without major environmental drawbacks? Of course, like most, I am concerned with the environment, but also, as a student, I am concerned with economic stability and employment opportunities for myself and for my peers and can only hope that many job opportunities will be open to me in the near future. If that means developing the North, shouldn't it be done?

Weick
I think one has to balance all these things. There is no single answer. People want to live in a variety of different ways. In the North some people may very well want careers, with wage employment, with industry, whatever. Many other people will want to remain as they are, pursuing the lifestyle that is essentially land based, and so on. Things never do work out ideally. In an ideal world, all these things could somehow be balanced. For example, Tom Berger mentioned the Cree hunter income security program. That program allows native people to participate in development without necessarily being in it themselves. They can remain hunters, but their participation is in the form of getting the necessary funding from government. The same kind of thing has been studied for the Northwest Territories. You could, for example, have major resource development, with native people not necessarily employed in that. On the other hand, you could have a lot of people working for industry, if that is what they want to do. It is not a question of either/or in any of these things.

Mag Johnstone (Edmonton, Alberta)
I have a question for Ambassador Rajakoski. We heard this morning about the failure of the educational system in Chesterfield Inlet. Some time ago I visited the Lapp Centre in Rovaniemi, Finland, and was thrilled at the way they were educating the northern Scandinavian peoples in the care of their reindeer herds and in living on the tundra. Is this school in operation at this time? That is my first question. The second question is: Is this type of education still being carried on—teaching the Lapps the skills necessary to live on the tundra and in their area? The third question is: Should this be the route and the pattern that Canadian schools should take, rather than having northern natives being taught a southern

curriculum and southern values by southern Canadian teachers? I would like your comments please.

Rajakoski
First of all, I hope that that centre school is still functioning. In fairly general terms I think that, as far as the aboriginal peoples are concerned, circumstances in the Arctic vary very much indeed. While I am no expert, I think that the situation in Finnish Lapland and Swedish and Norwegian Lapland is such that we do not have great problems. Development has been going smoothly for centuries, if not for thousands of years. And all the time we have had a relationship between the Southerners, as they call us, the majority of the Finnish people, and the Lapps. I know that quite recently, let's say during the last twenty or thirty years, much has been done to promote the Sami culture and language and to satisfy the local interests of the Lapps. I am not one to come and say what would be the correct formula in northern Canada, because I don't know those areas at all, but the development has been fairly smooth in the Nordic countries.

Emlen T. Littell (Sierra Club of Western Canada, Victoria, British Columbia)
Granted that local developments can be adequately regulated—though that does not seem to be a minor problem—what steps are necessary to reduce the importation of outside pollutants into the Arctic? This is a multinational problem of air and water degradation, as the pollutants emanate, as I have heard through the Canadian Arctic Resources Committee, from many nations. Both polar and nonpolar countries are just pouring pollutants into the Arctic. What can be done about it, Mr. Weick?

Weick
You are sure that question is for me? I am really not an expert in this area. All I can say is you need international cooperation. Maybe one of the other panelists would care to elaborate on that. I know it is happening, I deplore it, I am scared of it, and I think a lot of us are, but frankly I have no answers for you.

Lewis
Why doesn't John Merritt take a crack as well?

Merritt
Clearly the problem does require international cooperation and initiatives. I am very pleased, as are, I am sure, other people here today, about what the Finnish government is attempting to get started. My worry is that, when it comes to Arctic issues, there is a sense that military problems are

the overriding concerns of national governments, and that, because of the superpower stand-off, there is no real incentive for the smaller countries to show diplomatic initiatives on any front. I think that is perhaps one of the more disappointing aspects of Canada's response to Mr. Gorbechev's Murmansk speech, which touched not only on military topics but on a variety of other things.

I think Finland, along with a number of the other middle powers, such as Norway, Sweden and Iceland, would be very welcoming of a great deal of Canadian effort and determination to insist on an ambitious agenda of civilian cooperation. Clearly the environment has to be the number one point on that agenda.

Bill Seidel (St. Albert, Alberta)
After ten years of moratorium, Mr. Berger, can you honestly tell us the decision has enhanced the hunting and trapping in the area, or is there more alcohol, drug abuse and suicide than there was during the time of the heavier industrial activity? I have a couple of points to make. A COPE[6] member at Tuk said to me, "Bill," he said, "We will never go back to hunting and trapping. We are doing too well now and it is probably more of a front for political reasons that we are putting up such a battle on it." Further on, in the Fort McPherson area, they go hunting because it is closer to the caribou herds and they catch them and shoot them. One of them said he had shot eighty-eight and, he said, "I only take the hearts." He said, "I leave the rest because it is too much to bring back."
I only wanted to point out that there are two sides to it. What do you think about the going back to hunting and fishing, and what do you think about the drug and alcohol abuse? Has it increased or not since the moratorium has started?

Berger
That is a question that almost incites me to another speech, but let me just put it this way. When I held hearings back in the 1970s, I heard people who testified, and I accepted the views they publicly expressed and that they obviously held with great conviction. I heard, in coffee shops, many of the points of view that you have expressed. Long-time non-native northerners would say in the coffee shops to me, "Well, that's not really what people are thinking." But my view has always been that you accept what people say when they are concerned enough to come out and speak their mind in public hearings.

The second thing is that everybody knows that the social pathology of northern communities is a matter of grave concern especially to northerners. The statistics on alcohol abuse and teenage suicide and so on are alarming. I am not in a position to say whether they are better or worse since 1977. I don't know the answer to that, but I can tell you that in Alaska, where in the 1970s they did proceed with a large-scale frontier project, the TransAlaska Pipeline, the figures relating to the kind of social pathology that you and I are concerned with—crime and violence and family abuse and alcohol—became much worse than they had been. But I think that it is a mistake to try to argue about this in the present setting, in a fragment of conversation, because these are profound questions that I didn't purport to address in my talk today. I know that northern native peoples are concerned about these things; I am content to let them express their own views and make their own choices regarding the maintenance of the subsistence economy and large-scale industrial development. Let them make their own choices and they will have to live with the consequences. That's why I would rather that someone like myself shouldn't be telling them what to do on these subjects.

Seidel
I only have one point. I think that the administration should enforce school attendance beyond grades two, three and four, as we do with whites, rather than running a double standard.

Berger
Could I just add one thing to what you said. The question of northern development is one that I have tried to put in the broadest context; that is where I think it belongs. As Ed and I have both said, simply to say that it is either one way of life or the other, that if you get a VCR you have opted somehow for the whole urban, industrial way of life, is altogether a mistake.

My conviction after three years in the Mackenzie Valley and two years in Alaska, hearing from thousands of witnesses, literally thousands, is that there will never be wage and salaried employment for all people in the North and that we are fooling them if we tell them that through megaprojects we are going to give them full-time wage and salaried employment of the type we expect here in the metropolitan regions of the country. I do not think it can be achieved. There should be more jobs for native northerners in government and in business. Native people should be in business, in entrepreneurial activities, where that is appropriate and possible. But let us not pretend—and I think it is a pretense—that we can reproduce in the Arctic the kind of wage and salaried employment we expect to enjoy here in the metropolitan regions of the country. That's why the subsistence economy will remain absolutely essential to the life of northerners.

Lewis

I must say that it shows what conducting public inquiries and being a judge and a private citizen practising law in Vancouver does to one's equanimity. Tom responds by being incited to a speech. I would have been incited to physical aggression.

Notes

1. The $100 billion Grand Canal proposal has attracted media attention for the last three decades. Its authors envisage diking James Bay, which would then become a freshwater lake; the water would be pumped south to the Great Lakes and the United States. (Editor)
2. The motivating concern was that of the Soviet Union over Norwegian hunting off Svalbard and American hunting on the Chukchi Sea. Canadian and Alaskan polar bear populations remained healthy. (Editor)
3. Arviat, on the west coast of Hudson Bay, was formerly known as Eskimo Point. (Editor)
4. "We are from Arviat."
5. "It's very nice to hear the Inuit language spoken."
6. Committee for Original Peoples' Entitlement, a Mackenzie Delta–Beaufort Sea native group, since replaced by the Inuvialuit Regional Council. (Editor)

The Human Foundation for Peace and Security in the Arctic

Gordon Robertson

CHAPTER 8
The Human Foundation for Peace and Security in the Arctic

GORDON ROBERTSON

The context in which this paper fits is provided, I think, by some of the comments made this morning by Mary Simon and Tom Berger.

Basically, the proposition you will hear me develop is that political development for the North, giving greater power and better protection to our native people, is very much in the national interest in relation to the development of an international Arctic policy by Canada. The other dimension of my proposition came out this morning in the comments that arose in the papers by Mary Simon and Tom Berger: that such a program would also be very much in the interests of the native people. You may remember that Mary Simon said that exclusion of native peoples from policy decisions on economic and political development was a major problem. And you heard that very touching episode when two Inuit questioners asked how they could safeguard their culture and their language when

Gordon Robertson is a fellow in residence and a former president of the Institute for Research on Public Policy in Ottawa. Educated at the universities of Saskatchewan and Toronto and at Oxford University, he joined the Department of External Affairs in 1941 and subsequently worked in the Prime Minister's Office and the Cabinet secretariat. He served as deputy minister of the Department of Northern Affairs and Natural Resources before his appointment as Clerk of the Privy Council in 1963. From 1968 until his retirement in 1979, as secretary to the Cabinet for federal-provincial relations, he participated in the constitutional review. In 1986 the institute published his Northern Provinces: a Mistaken Goal. *Mr. Robertson is a member of the Queen's Privy Council for Canada and a Companion of the Order of Canada.*

they have "no economic or political power"; and Mary Simon said that they needed not only economic but political development in order to do that.

Later on, Tom Berger was dealing with the question of economic development versus maintenance of subsistence on the land, that is, the lifestyle of the native people, and he commented again on the importance of the native people being in a position to make their own decisions on the mix of issues that is involved in that kind of question. And, again, I think that political development is the answer to the problem. So that is, in a preparatory way, the context in which my comments fit.

[Editor's note: Passages in italics were in the written speech but were not delivered orally because of time constraints.]

There has been a tendency in our national attention—or inattention—to the North to oscillate between apathy and excitement; too little and too much in interest and enthusiasm, in hopes and in disappointments. Prime Minister St. Laurent said in 1954 that Canada had, up to that time, administered the North "in a state of absence of mind": few had cared about it and little had been done. Prime Minister Diefenbaker, four years later, lit up Canadian politics and aroused national fervour with "The Vision"—a resource-rich North on which our future greatness would depend. The picture was too glowing and disillusionment followed.

The risk is still with us. It would be superficially attractive to suggest an "ocean-basin" focus in world history: the Mediterranean in the ancient world; the Atlantic Ocean and its bordering countries during the last four hundred years; the Pacific Rim today and in the coming century. One could move on to envision a future focus on the Arctic Ocean for the twenty-first century. That would be provocative and stimulating. It would also be a mistake. The Arctic is not going to be that central or that important. But it is not a mistake to believe that the international Arctic is going to have much more importance in the future than it has had in the past. Nor is it wrong to believe, as the special joint committee of the Senate and the House of Commons on Canada's international relations did in 1986, that there should be "a northern dimension for Canadian foreign policy." After all, a large part of the Arctic is ours, and it is the area above all others where Canada can and should play a significant role. The question left unanswered in 1986, and not at all clarified by the government of Canada since, is what the object of Canadian policy for the Arctic should be.

In 1988 a working group of northern specialists, which I chaired, published a report, *The North and Canada's International Relations*. The working group recommended that the aim of Canadian policy should be "the achievement and maintenance of a secure and peaceful world in the Arctic." It went on to suggest that that peaceful world should be one "in which aboriginal inhabitants can preserve the essentials of their cultures while living in association with Canadians of other origins."[1]

I draw attention to that part of the working group's recommendations because it would be very easy to have a northern dimension for Canadian foreign policy that took little or no account of the interests of the Inuit people. They are very few in Canada, only 25,390 according to the 1981 census—one-tenth of one percent of our population. Just enough to make a good-sized Alberta town. The objectives of Canadian Arctic policy, if set in broad national terms of politics, economics, defence, science or environment, could very easily find little or no scope for Inuit concerns. Canada could pursue peace and security in the Arctic as if it were an uninhabited world, which much of it is. But if we did, we would fail in our moral obligation to the Inuit people. And we would also end up, I would submit, with a hollow policy; inadequate, without credibility internationally and doomed to fail in the long run for lack of any sound base in Canada's domestic North.

I suggest to you that we cannot have a successful policy for the international Arctic if we do not have a population in the Canadian Arctic that is self-reliant, self-respecting and self-directing. That population will be largely Inuit apart from itinerant specialists from the South. Most other Canadians will not choose to make their homes in the Arctic. The Inuit are a people who have been confident and vigorous in the past. However, the transition in the last thirty years to a new way of life and the impact of modern culture on the Inuit have produced social and economic problems that threaten their traditional self-reliance. As we heard this morning, they threaten too the pride and confidence of the Inuit in their own identity.

We have seen that kind of erosion of self-reliance and self-respect before in our history. It has happened with some of the Indian bands in the provinces of Canada that have found the economic and social problems of their lives today insoluble. We have seen the results in family disintegration, alcohol abuse, unemployment, juvenile delinquency and other conditions that frustrate and humiliate us all as Canadians. Tom Berger referred to this as social pathology, an excellent phrase. Such erosion has not yet gone nearly as far with the Inuit, but it could easily do so if we do not find solutions to emerging problems. We will not be very convincing in any international Arctic policy we may devise or in our relations with other Arctic countries and Arctic peoples if we do not have domestic Arctic policies that produce and maintain in the North a vigorous, self-governing, successful Inuit community.

Domestic policies of that kind will require more imagination, more initiative and more commitment by our federal government than it has demonstrated so far. Action is required on both the political and the economic levels. The Inuit of the Northwest Territories, who are the vast majority of Inuit, have defined their political objective; they did it ten years ago when the Inuit Tapirisat gave formal support to the establishment of a new territory of Nunavut. It would comprise that part of the Northwest Territories north and east of the tree line. The Inuit would constitute about eighty percent of the population of the new territory, but they have made it clear that they contemplate a system of government for all people in the territory, Inuit and others. Our federal Liberal government, from the time of that resolution in 1979 until 1984, and the present Conservative government, since 1984, have smiled benevolently on the idea of a new territory of Nunavut, but they have done nothing to help bring it about.

The politically comfortable course for the federal government is to stay safely away from any position on the issues at stake. That is largely what it has done. This is a short-sighted policy, if it can be called a policy at all. It fails to give weight to the future consequences of effortless drift today—drift that finds an excuse for inaction in local difficulties in reaching agreement on a boundary for the new territory. This policy is not only short-sighted, but also, in my judgment, irresponsible.

The present comfortable inaction means not doing what only our national government can do; that is, assessing what the national interests are in having a political structure in the Canadian Arctic that responds to Inuit aspirations and provides for Inuit self-government. This is not simply a territorial issue. There is a national dimension, which is what our national government is supposed to look after. Unless that is realized and unless there is a policy that responds to national concerns, a great opportunity may be missed. If it is, we will pay for it in the years ahead.

There is little point in trying to attribute blame for our past and present failure to find a successful solution for the economic and social policies of our aboriginal people. The failure is an unhappy fact. A good deal of progress is now being made on land claims. That is encouraging but it is only partial. It does very little for the problem of aboriginal self-government, on which we saw four constitutional conferences end in failure in 1987. There are complex reasons why a solution to the problem of native self-government is difficult in the provinces. There are no comparable reasons why it should be difficult in the Northwest Territories, where there are no provinces and where the native people constitute a majority of the population.

Within the Northwest Territories the position of the Inuit people, apart from the Inuvialuit in the West, has been clear for some time. In a plebiscite in 1982 more than seventy percent of eligible voters in the Eastern Arctic voted, and they voted more than four to one in favour of a territory of Nunavut. In the Northwest Territories as a whole, fifty-six percent of the vote was in favour of dividing the Territories in two. The precise boundary was, as I have indicated, the main problem. The legislative assembly of the Northwest Territories set up three entities to work on division: the Nunavut Constitutional Forum, the Western Constitutional Forum, and a joint entity called the Northwest Territories Constitutional Alliance. In January 1985, the Alliance agreed in principle on a solution to the boundary problem and, it seemed, to the difficulties of the Inuvialuit. The agreement fell apart. In January 1987, just over two years ago, a new agreement was reached at Iqaluit. That in turn has been held up by unresolved boundary disputes. They relate primarily to the definition of Indian and Inuit areas of land use.

Agreement obviously must be sought among the native people whose land use is at issue. But never have we been so close to a solution that would permit innovative systems of government to be introduced in Canada that would reflect the values of native people and provide future protection for their languages and cultures. The Dene and Métis in the western part of the Northwest Territories are as concerned about these as are the Inuit in the east and the Inuvialuit in the northwest. It is difficult to avoid the suspicion that ten years after the resolution of the Inuit Tapirisat and seven years after the plebiscite in the Northwest Territories no solution has been found for new political structures in the Northwest Territories because the present governments, both federal and territorial, are not interested in finding one. The homely adage about where there is a will there is a way is applicable. The will, I suspect, is lacking.

To return to my basic thesis: all of this question about self-government for the Inuit—and for the Dene and Métis in the western Arctic—is important if we are serious about Canada playing the role it ought to play in policies and international relations in the Arctic.

Apart from the direct defence issues and the related East-West confrontation in the Arctic, with virtually every question that will arise in our relations with other Arctic countries we will find our position stronger and more credible if we have a thriving, self-governing Inuit people in the Arctic. Our argument for sovereignty over the waters of the Canadian Arctic Archipelago will be much more persuasive to other countries and in international forums if that sovereignty is clearly needed to protect the environment and the economic base of a native population that uses the resources of the land. That land includes the waters of the Archipelago even more than the dry land of the islands that make it up. Our sovereignty is important if it is for a real purpose: for the welfare of real people who inhabit the Arctic. It will become largely empty symbolism if there is no self-reliant Arctic population there to benefit from our sovereignty.

Our relations with Greenland, where government is now in the hands of Greenlanders, who are basically Inuit, will be on a different basis if we have a government in Canada that can speak for an Arctic population that is eighty percent Inuit. Our discussions with the five Nordic countries will become more substantive and more credible if there are governments in the Canadian North that are a demonstration of our concern about principles underlying new political institutions now being set up in Norway in the interests of the Sami people, who are the inhabitants of the Norwegian Arctic. They, the Norwegians, are taking the aboriginal rights issue seriously. The Norwegian Sami Rights Committee in its report of 1984 recommended an addition to the Norwegian constitution to impose an obligation on the government "to enable the Sami population to safeguard and develop their language, their culture and their social life." The constitutional amendment has, I believe, passed the Norwegian parliament, and an elected Sami parliament is now being created. How do we look if we have done nothing about aboriginal self-government in the Canadian Arctic, where we have had efforts by the Inuit people for years?

The theme of this meeting is "The Arctic: Choices for Peace and Security." Peace and security in the Arctic are important for all Canadians and for the world generally. For whom are they more important than for the native people who live north of the Arctic Circle? They are the ones our government should have most in mind. Whatever we in Canada do towards the achievement of the objective we are discussing today will be less likely to succeed and of less purpose if it is not based on and in the interests of a proud, confident and self-governing people in the Canadian Arctic.

Notes

1. The report had in mind the Inuit in particular, since they are the natural, aboriginal inhabitants of the true Arctic. Moreover, their Arctic is the global Arctic, virtually all the inhabited world north of the Arctic Circle. The Inuit of Canada share the culture, language and many of the interests of the Inuit of Alaska, Greenland and the Soviet Union.

CHAPTER 9
International Arctic Cooperation: A Canadian Dream or a Necessity?

WALTER SLIPCHENKO

I will start with a quotation from Franklyn Griffiths and Oran Young:

> The ice is beginning to move under our feet. To those who live in the region and to knowledgeable persons elsewhere, there is a sharpened sense of the need to act. To be sure, the Arctic has not been immune to change in the past. But the rate of change today is new. And whether the issue is military, strategic, political-economic, environmental, or some combination of all of these, Arctic developments seem to be moving in directions that few really want. New ways of thinking about the region are required if national and international policies are to meet the gathering challenge of change.[1]

Many of the changes that have occurred during the last thirty years were brought about by activities in foreign regions including the foreign North. All of these activities are having increasing influence on developments in the Canadian North. I refer specifically to the following:
• Resource exploitation and industrial development in foreign circumpolar regions with their attendant environmental disturbances and impacts
• Atmospheric and oceanic pollution, including climatic change because of worldwide pollution
• The depletion of marine stocks
• International social, economic and political developments, including greater native participation in government
• The formation of international native organizations, such as the Inuit Circumpolar Conference, the World Council of Indigenous People and Indigenous Survival International

International Arctic Cooperation: Canadian Dream or Necessity?

Walter Slipchenko

Specifically, if you look at the developments taking place, you will find that Greenland, with its home rule government and its devolution of powers, is having an effect on what is happening in other parts of the world. In Alaska you have the implementation of the Alaska Native Claims Settlement Act, the movement toward tribal sovereignty—which affects native people throughout the world—the recent U.S. Arctic science policy initiatives, and you heard this morning about the International Porcupine Caribou Agreement and the possible developments in the Arctic National Wildlife Range. In the Soviet Union there is the offshore oil and gas development and the possibility of sales of offshore equipment; and there is air pollution occurring in a number of its large industrial centres. Of course, this air pollution is occurring all over the world.

The results of these influences have certain positive aspects, such as increased circumpolar interaction among native people, among northern regional governments and among scientists, but the influences, because of industrial development, have primarily been negative. However, spinoffs have included, interestingly enough, recognition of Canadian northern activities and expertise by other polar and nonpolar nations. In particular, the foreign North has developed interest in Canada's accumulation of knowledge and experience dealing with the political development of native people, the protection of the environment, northern construction techniques and the application of cold region technology in oil and gas offshore exploration.

As a result of international and national developments, influences, interactions and other happenings, such as the voyage of the *Polar Sea* through the Northwest Passage, a special joint committee of the Senate and the House of Commons produced a report in 1986 to help chart an independent Canadian foreign policy. The report underlined that "the Arctic region is rapidly becoming an area of international attention. Canada's huge stake in this region requires the development of a coherent Arctic policy, an essential element of which must be a northern dimension for Canadian foreign policy."[2] We are all waiting with bated breath to see

Walter Slipchenko is director of Circumpolar Affairs at the Ottawa liaison office of the government of the Northwest Territories. He was educated at Royal Roads and Royal Military College, and at Queen's University and the University of Manitoba. In 1966 he joined the Department of Indian Affairs and Northern Development in Ottawa, as a researcher on the Soviet North, and subsequently established an office concerned with circumpolar relations in the department. He has maintained an active role in the negotiation of scientific relations with the Soviet Union. He accepted his present position in 1988.

when we get this northern Canadian foreign policy.

The committee also argued that it was essential for Canada to collaborate bilaterally or multilaterally with all northern states.[3] Joe Clark, Secretary of State for External Affairs, responded for the government by noting the importance of promoting enhanced circumpolar cooperation as an important part of an "integrated and comprehensive northern policy."[4] Furthermore, in the section dealing with "A Northern Dimension for Canadian Foreign Policy," Mr. Clark agreed with the majority of the Committee's recommendations, which dealt primarily with an expansion of cooperative arrangements with all northern countries.[5] Almost one year later, in the Soviet Union, a similar decision dealing with enhancing circumpolar cooperation was reached, when, on October 1, 1987, General Secretary Mikhail Gorbachev outlined six proposals for Arctic bilateral and multilateral cooperation.[6]

It is within this framework that I would like to discuss the need for international Arctic cooperation, past, present and future. The discussion involves the following:
- The historical setting of international Arctic science cooperation
- The types of international Arctic cooperation agreements
- The proposed international Arctic science committee
- The Canada–U.S.S.R. Arctic Science Exchange Program
- The implications of Mr. Gorbachev's speech

International Arctic science cooperation extends back a hundred years to the creation in 1879 of the International Polar Commission, which led to the International Polar Year of 1882–83. Eleven nations (including all polar nations) sponsored fourteen polar research expeditions. In 1932–33 the second International Polar Year followed by studying Arctic phenomena, and in 1957–58 the International Geophysical Year was initiated, with scientific investigations in both polar regions. In 1958, the Scientific Committee on Antarctic Research (SCAR) was created as a scientific committee of the International Council of Scientific Unions (ICSU).[7]

Since then, according to Dr. Fred Roots,[8] a series of different types of scientific activities with Arctic components has taken place. These scientific activities can be grouped under seven extended categories:

1. Large, coordinated regional or circumpolar single programs, such as the polar segment of the international magnetosphere study
2. Arctic components of complex and highly organized worldwide research programs, such as POLEX (part of the Global Atmospheric Research Program)
3. Scientifically focused wide-ranging programs to explore new concepts, gather data or compare phenomena, such as the Arctic Ice Dynamics Joint Experiment conducted in the Beaufort Sea (1969–1976)
4. Programs for collaboration by a number of researchers or agencies in

one area or where taking advantage of an outstanding opportunity will benefit many others

5. Studies that are so sophisticated and so expensive that no one country can provide the expertise or facilities, so that international collaboration is essential, such as the Marginal Ice Zone Experiment and the Greenland Ice Sheet Program

6. Programs designed to facilitate the sharing and exchange of experience and knowledge or to compare results of similar studies, such as the 1984 Canada–U.S.S.R. Arctic Science Exchange Program

7. Programs designed to combine data or information from several sources into a common format or synthesis that can be useful to many countries, such as the bathymetric map of the Arctic Ocean produced in the General Bathymetric Chart of the Oceans series[9]

The results of this research indicate the following:

• Arctic nations share many common problems that require cooperation in research and in the sharing of scientific data

• Arctic research is difficult and costly, but the results are vitally important nationally and internationally, particularly to northern people

• Arctic science requires the financial and scientific support not only of polar but also of nonpolar nations

• Environmental and social concerns have emerged as among the most important Arctic issues today

• Social and political developments in the Arctic have altered the demand for knowledge and exchange of information[10]

In other words, circumpolar scientific research is an essential component of international cooperation, which will continue to grow rather than diminish as polar nations attempt to resolve complex northern problems.

There are basically two types of cooperation that take place in the Arctic. One is multilateral and the other is bilateral. There are three multilateral agreements or regimes that prevail in the Arctic, one of which, the polar bear agreement, is totally multilateral in that it affects all of the northern polar regions.[11] The agreements are as follows:

• The Svalbard Treaty, which was signed by forty signatories (it is limited to a small area)

• The International Agreement on the Conservation of Polar Bears, which operates within the framework of the International Union of the Conservation of Nature and Natural Resources; the signatories include Canada, Denmark/Greenland, Norway, the Soviet Union and the United States

• The 1911 Treaty for Preservation and Protection of Fur Seals, which was replaced by the interim Convention for the Conservation of North Pacific Fur Seals, which was extended and amended in 1963, 1969, 1976, 1980 and 1984; the 1984 protocol was not ratified by the United States

In addition, a number of international scientific organizations, such as the International Union of Circumpolar Health, the International Permafrost Association, the Comité Arctique and the Arctic Ocean Science Board, are multilateral. Their prime responsibilities are to hold conferences, provide communication and exchange information.

Unfortunately, there is no international body for the coordination of circumpolar scientific activities similar to the Scientific Committee on Antarctic Research. However—and I believe it was mentioned this morning—discussions have taken place in the last two years that may result in the formation of an international Arctic science committee, which "would serve the scientific interests of arctic countries and provide a forum for discussion and coordination on the research interests of any country involved in arctic science."[12]

Founding articles for this committee were drafted in Leningrad, on December 10 and 11, 1988, at a meeting attended by representatives of all eight Arctic countries (the Soviet Union, the United States, Denmark/Greenland, Iceland, Norway, Sweden, Finland and Canada) and this draft is now being reviewed by each of the countries. If all the representatives agree, the next stage will be the formation of such a body. It will be a nongovernmental scientific organization, the prime purpose of which will be to encourage and facilitate international scientific cooperation[13] (though in practice there undoubtedly will be varying governmental involvement). The lack of a multilateral circumpolar arrangement has created difficulties that have been offset in part by bilateral arrangements.

A number of agreements and programs focusing on Arctic science or on some aspect of the Arctic do exist between countries. There are several such agreements dealing with environmental, cultural and scientific arrangements, particularly between the Soviet Union and other polar countries.[14] Canada at the present time is involved in three such bilateral arrangements:

1. The 1983 agreement between the government of Canada and the government of Denmark (the Marine Environmental Cooperation Agreement), which covers ships and offshore pollution and under which research in other areas is possible.

2. The Canada–United States Agreement on Arctic Cooperation, which generally facilitates navigation by icebreakers in Arctic waters, and which provides for the sharing of research information in order to enhance the understanding of the marine environment.

3. A protocol of Canadian-Soviet consultations on the development of a program of scientific and technical cooperation in the Arctic and in the North, which was signed in 1984 and renegotiated on February 26, 1987, and which has become known as the Canada–U.S.S.R. Arctic Science Exchange Program. It provides for the development of a program of sci-

entific and technical cooperation in the Arctic and the North.

I would like to speak about the Canada–U.S.S.R. Arctic Science Exchange Program, as I feel that it has demonstrated real northern cooperation between Canada and the Soviet Union.

Interest in the possibility of development of Arctic cooperation with the Soviet government was first mentioned in October 1955, when Lester B. Pearson, as Secretary of State for External Affairs, visited the Soviet Union.[15] In 1959 Prime Minister Diefenbaker stated that the Canadian government had initiated discussions with the Soviet government on northern research and administration.[16]

It was, however, the visit to the Soviet Union of two ministers responsible for northern affairs, Arthur Laing in 1965 and Jean Chrétien in 1972, that marked the real beginning of discussions and activities to initiate an effective exchange between the two countries.[17] I must say that all subsequent ministers responsible for northern affairs, be they Liberal or Conservative, have always supported this exchange program.

During the 1970s, although both the Canadian and Soviet governments had selected the Arctic as the region that held the most promise for bilateral cooperation, and in spite of several joint discussions and the signing of two memoranda dealing with earth sciences, meteorological and oceanographic studies, marine and terrestrial ecosystems and medical sciences, the two sides could not agree upon a mutually acceptable program. The basic problem was that the Canadian side argued that any exchange in the North must include "people's issues," that is, social sciences, while the Soviet side insisted that the program initially should be limited to the physical and biological sciences. The result of this was a stalemate; we agreed to disagree. Other events, such as the invasion of Afghanistan, overtook the discussions and nothing happened until 1981, when the Soviet side proposed the inclusion of social sciences as a valid subject in the exchange program.

Further discussions were held, many of which suffered from the syndrome I call "one step forward, two steps back." Finally a protocol was signed in Moscow in April 1984 in which the two sides agreed to a detailed two-year program of activities in eighteen main subject areas under four major themes: geoscience and Arctic petroleum; northern environment; northern construction; and ethnography and education.

During the last five years, a number of Canadian and Soviet delegations have been exchanged, involving over 150 Canadian and Soviet scientists and specialists. Although the first exchanges were primarily familiarization visits, the last years have generally been very productive, with such overall benefits to the Canadian side as closer personal ties between specialists; development of good working relationships; publication of a number of reports; access to information that had not been available previously; firsthand knowledge of the situation in northern areas, particularly of the state of the art of Soviet technology; teacher exchange; possible commercial spinoffs; and the involvement of specialists from the regions, including the participation of aboriginal people.

On the Canadian side, there is active support for this exchange both inside and outside government, with good representation from the federal and territorial governments.[18] In fact, as a result of this program, the Northwest Territories has established direct links in northern matters not only with the Soviet Union but also with other circumpolar nations.[19] The program has also been instrumental in helping to bring about the Quebec–U.S.S.R. protocol dealing with northern matters. Moreover there can be no doubt that this exchange helped to improve Canadian-Soviet relations in general.

It should also be noted that, since this exchange was "the only game in town" from 1984 to 1989, various types of exchange were encouraged under this umbrella whether or not they were strictly scientific. It became obvious that we needed a broader agreement. As a result, following the Soviet initiative for a cooperation agreement in Arctic science in 1987 and as a result of northern interests being incorporated, we now have, if it ever gets signed, a mutually beneficial all-purpose agreement with the Soviet Union dealing with Arctic cooperation.

There appears to be a tentative understanding by both governments to sign this agreement, which has been agreed to in principle by both countries. It would incorporate the present Canada–U.S.S.R. Arctic Science Exchange Program and programs, which have been developed in other areas such as economic cooperation; social, cultural and economic development of native people; and cultural and academic exchanges.[20]

I now would like to discuss the results of the initiatives for the Arctic that were raised in Mr. Gorbachev's speech of October 1, 1987. He outlined a plan for Arctic cooperation, which included a nuclear-free zone in northern Europe; restriction of naval activity in northern waters; cooperation in the development of resources in the Arctic; study of the feasibility of a joint Arctic scientific council; cooperation for environmental protection; and the possibility of opening up a northern sea route from Europe to the Far East.

[Editor's note: Because of time constraints, Mr. Slipchenko delivered a very brief summary of the italicized passage which follows. It is included here in its entirety for the additional information it contains.]

As a result of Mr. Gorbachev's speech, international cooperation in resource development will increase the ongoing joint ventures in the Arctic. Canada and Norway were singled out in the speech because of their offshore expertise. Canadian activities during the 1970s were initiated under the 1971 Canada–U.S.S.R. Industrial Application of Sci-

ence and Technology Agreement (INDEXAG), resulting in some equipment sales in oil and gas, testing of Soviet turbo drills in Canada and testing of Canadian technology in the Soviet North.

During the 1980s Lavalin Inc. gained contracts in excess of $300 million in the southern region of the Soviet Union (Volga/Ural) and is now looking at projects farther north. The firm Foremost has been successful, having sold their first special-purpose Northern Swamp Vehicle in the early 1970s, before the inception of any agreement, and became a party to the first Canadian-Soviet venture in 1985 in the joint development of a (seventy-ton) all-terrain, tracked vehicle, the Yamal.[21]

Current onshore and offshore northern development activities in the Soviet North by Canadian and foreign firms have increased in the last year.[22] *Canadian involvement remains small in comparison with the activities of other countries, although the provision in 1988 of a $500 million line of credit by the Canadian government for Soviet buyers of Canadian products will help stimulate joint ventures and some purchases of Canadian equipment.*

The proposed international Arctic science committee will be supported by the Soviet Union, if the highly successful Arctic Science Conference, held in Leningrad during December 1988, is any indication of interest in Arctic cooperation. The call for environmental protection in the Arctic is, of course, more significant to Canada and to other polar countries. Soviet cooperation will help in implementing the recommendations of the Brundtland Commission providing opportunities for circumpolar cooperation in an area most important to the survival of the North.

I would like to conclude by emphasizing the need for international Arctic cooperation and for Canada to take the lead role as an "honest northern broker" to achieve that end.

Addendum

Mr. Slipchenko has provided the following list of arrangements and programs that polar countries have with the Soviet Union, as categorized by G. Osherenko.[23]

Environment and Conservation Arrangements

1957	Norway/U.S.S.R. Agreement on Measures to Regulate Sealing and to Protect Seal Stocks in the Northeastern Part of the Atlantic Ocean
1972	U.S./U.S.S.R. Agreement on Cooperation in the Field of Environmental Protection (Extended)
1973	U.S./U.S.S.R. Agreement on Cooperation in Studies of the World Ocean

1976	U.S./U.S.S.R. Convention Concerning the Conservation of Migratory Birds and Their Environment
1984	Canada/U.S.S.R. Protocol on the Development of a Program of Scientific and Technical Cooperation in the Arctic and the North
1988	Norway/U.S.S.R. Agreement Concerning Scientific-Technological Cooperation on Problems of the Study of the Arctic and of the North for 1988–1992 (See section IV, Arctic Biology)
1988	Norway/U.S.S.R. Agreement on Environmental Protection

Industry and Commerce Arrangements

1956	Norway/U.S.S.R. Search and Rescue Cooperation in the Barents Sea
1957/1973	Norway/U.S.S.R. Sea Boundary Demarcation Agreements

Culture and Science Arrangements

1977	U.S./U.S.S.R. Soviet Academy of Sciences (Institute of Ethnography)/Smithsonian Institute Joint Exhibit "Crossroads of Continents: Traditional Cultures and Peoples of the North Pacific Rim"
1985	U.S./U.S.S.R. General Agreement on Contacts, Exchanges and Cooperation in Scientific, Technical, Educational, Cultural and other Fields
1985	U.S./U.S.S.R. Program of Cooperation and Exchanges for 1986–1988
1984/1987	Canada/U.S.S.R. Protocol of Consultations on the Development of a Program of Scientific and Technical Cooperation in the Arctic and the North
1988	Norway/U.S.S.R. Agreement Concerning Scientific-Technological Cooperation on Problems of the Study of the Arctic and of the North for 1988–1992

Notes

1. Franklyn Griffiths and Oran R. Young, "Impressions of the Co-Chairs," Reports and Papers 1988-1, Working Group on Arctic International Relations, First Session (Hveragerdi, Iceland, July 20–22, 1988): 1.

2. *Independence and Internationalism,* a report of the Special Joint Committee of the Senate and the House of Commons on Canada's International Relations (Ottawa, 1986),

3. Ibid., p. 130.

4. *Canada's International Relations: Response of the Government of Canada to the Report of the Special Joint Committee of the Senate and the House of Commons* (Ottawa: Secretary of State for External Affairs, 1986): 31–33.

5. Ibid., pp. 85–87.

6. Speech by Mikhail Gorbachev, General Secretary of the CPSO Central Committee, at the presentation of the Order of Lenin and Gold Medal in Murmansk, October 1, 1987. English translation provided by FBIS-SOV-87-191, October 2, 1987, pp. 41–42.

7. E. F. Roots, "Cooperation in Arctic Science—Background and Prospects." Keynote address to the meeting on International Cooperation in Arctic Science, the Royal Swedish Academy of Sciences (Stockholm, Sweden, March 24–26, 1988): 3. This paper is an excellent review of developments in Arctic science and outlines the directions that multilateral scientific cooperation should take.

8. Science Advisor to the Department of the Environment, Canada.

9. E. F. Roots, "International and Regional Cooperation in Arctic Science: A Changing Situation," *Muskox,* No. 14 (1986): 20–21.

10. See Roots (1988), op. cit. pp. 1–8. This information was summarized from statements made by Dr. Roots.

11. Oran R. Young, "International Cooperation in the Arctic: Past Experience and Emerging Opportunities," Science For Peace Conference on Arctic Cooperation (Toronto, October 26–28, 1988): 11–15.

12. Oran Young, op. cit., p. 19.

13. The fourth proposal in Mr. Gorbachev's Murmansk speech supports establishment of a joint Arctic council; the proposed international Arctic science committee may fit that need.

14. Gail Osherenko, "Environmental Security in the Arctic: Prospects of Soviet Cooperation," draft paper (November 23, 1988). See the addendum to this address.

15. See John Hanigan, "New Dimension in Canadian-Soviet Arctic Relations," *Point of View,* Canadian Institute for International Peace and Security (November 1988): 2.

16. Ibid. The author reviews Canadian-Soviet relations in the Arctic and argues for the necessity of a northern foreign policy. Moreover, he outlines the recent developments by the government of the Northwest Territories in expanding relations not only with the Soviet Union but with other circumpolar countries.

17. Walter Slipchenko, "Canada–U.S.S.R. Arctic Science Exchange Program: An Historical Perspective of Cooperation in the Arctic," paper presented at the Soviet Maritime Arctic Workshop, Woods Hole Oceanographic Institute, May 10–13, 1987.

18. This exchange program has been supported in the report of the special joint committee of the Senate and the House of Commons on Canada's international relations (op. cit., pp. 129–130) and in the government's response (op. cit., p. 86).

19. Government of the Northwest Territories, *Direction for the 1990s.*

20. In addition to the incorporation of northern interests in the proposed agreement, consultations were held with the governments of Yukon and the Northwest Territories and with native associations to ensure that northern needs would be looked after. It is unfortunate that such an extensive delay has occurred in signing the agreement. This delay has resulted in certain programs being held in limbo, as there is a general ignorance of how the various parts will fit into the new mosaic. Nevertheless, once signed, this agreement will contribute significantly to the expansion of polar endeavours with the Soviet Union.

21. Carl H. McMillan, "Joint Ventures in Arctic Resource Development," *Northern Perspectives,* 16, 4 (July-August 1988). In this article, the author reviews Canada's past performance under INDEXAG and agrees that opportunities for Canadian firms are present; however, they will not be easy to develop.

22. Current Canadian-Soviet arrangements during 1988 included: (1) a proposed Soviet

joint venture with Canadian Fracmaster Ltd. to raise the output of low-yielding oil deposits in Siberia; (2) collaboration between Falconbridge and Monter (Yugoslavia) to build a hospital in Norilsk; (3) the participation of twelve Canadian firms, under the auspices of the Ministry of Economic Affairs of Alberta, in discussions in Moscow concerning offshore and onshore activities in the Barents and Okhotsk Seas; and (4) talks between Gulf Canada and the Soviets on possible cooperation in the Arctic.

23. See footnote 14 and associated text.

CHAPTER 10
The Role of International Law for Peace and Security in the Arctic

DONAT PHARAND

I was asked to speak on the role of international law with respect to peace and security in the Arctic. Let me say first of all that it is not possible to divorce the role of international law with respect to peace and security in the world generally from its role in the Arctic. It is the same. Its primary role is, of course, to have certain fundamental norms for the conduct of states, so that those states will be able to live, regardless of their size, in peace and security.[1]

This means, perhaps unfortunately, that we also have to speak about sovereignty. You are going to ask me why. Well, we still live in a world of sovereign states. It is very unfortunate. I would be the first one to have a worldwide federation, but let us be just moderately realistic. It is not for tomorrow. States still insist on their political independence and sovereignty and are very sensitive; they say immediately, "If you touch my sovereignty my security is at stake." So I have no choice but to speak about sovereignty.

On the other hand, we do live in an interdependent world. It is getting smaller all the time. Of course, our moderator knows that better than anyone else, having just spent four or five years at the United Nations. We must therefore cooperate. So I will speak, Mr. Moderator, on these three points: security, sovereignty and cooperation, strange as these three sides of the same coin might sound.

International law and security in the Arctic: What are the basic legal principles? They are very simple. There are two fundamental legal principles. They are laid down in the charter of the United Nations. The first one is that there is an obligation for all states to settle their international disputes by peaceful means; of course, the charter goes on and specifies a

105

The Role of International Law for Peace and Security in the Arctic

Donat Pharand

number of those means. That is not too difficult, though I do not mean that it is well implemented.

The second principle is more difficult. It is the abstention from the use of force and indeed from the mere threat of the use of force. Now this is completely new. Before 1945 this was not in existence as a principle. Indeed in the covenant of the League of Nations, war was still permissible. If the Council of the League was not unanimous in its recommendation with respect to the settlement of a dispute, parties reserved to themselves the right, in the words of the covenant, to take whatever action they saw fit after a cooling off period of three months. Can you imagine? War was still permitted. Then in 1928 we had the General Treaty for the Renunciation of War, the Briand-Kellogg Pact. The only thing it did was to prohibit war as a matter of national policy. Can you imagine?

So we made, theoretically speaking, a lot of progress when in 1945 we agreed in San Francisco to insert in the charter of the United Nations, as perhaps the most fundamental principle, abstention from the use of force and the threat thereof. This means, of course, that war, theoretically speaking, is completely abolished—even the use of force and, I repeat, the threat of force.[2] However, this is theory. In practice, you know as well as I do that it is not quite the case.

And we have nuclear weapons. That is my second point. Are nuclear weapons legal in international law?[3] I think this is a legitimate question to ask. Would you believe that I have to answer in the affirmative to that question? It is abominable that I should have to answer in the affirmative, but that is my humble opinion.

I think that the consensus among international lawyers is unfortunately that the only prohibition we have, strictly speaking, is contained in the Nonproliferation Treaty of 1968 and that prohibition applies only to the non-nuclear-weapons states.[4] Do not forget that this was an agreement between the nuclear-weapons powers, but only three of them, because the other two didn't come in—the United States, the United Kingdom and the Soviet Union are parties, but France and China are not. It was an agreement between them not to transfer nuclear-weapons technology to others,

Donat Pharand is a professor emeritus of the University of Ottawa and a consultant in international law. He holds degrees in law from Dalhousie University, the universities of Paris and Michigan, and from The Hague Academy. He served in the Canadian army and air force from 1943 to 1946 and taught for nearly thirty years in common law, civil law and political science. He is an authority and author on the Law of the Sea and the status of Canadian Arctic waters. He is a Queen's Counsel, a Fellow of the Royal Society of Canada, and a Fellow of Clare Hall, Cambridge University.

and an agreement by the non-nuclear-weapons states not to receive it.[5] That is the whole extent of the treaty. Do not forget that it applies only to the development and the use of nuclear weapons and nuclear explosive devices. It does not apply to nuclear nonexplosive devices: it does not apply—that treaty, I repeat, does not apply—to their nonpeaceful uses, and it does not apply to nonproscribed military activities.

What am I talking about with this legal jargon? Do you know that this legal jargon, which is used in Article 3 of the Nonproliferation Treaty and in Article 14 of the agreement between Canada and the United Nations International Atomic Energy Agency, justifies any non-nuclear-weapons state such as Canada in having weapons-grade nuclear materials?[6] Canada is permitted therefore to use nuclear material for the propulsion of ships, including the propulsion of submarines. This is how Canada could legally justify proposing to acquire ten to twelve nuclear-powered submarines in the Arctic.[7] Whether it can justify it politically is another matter. I am not saying that the answer should be in the affirmative, not at all.[8]

But I am speaking about sovereignty.[9] Why should Canada acquire those nuclear-powered submarines? To protect our sovereignty, we were told, mainly in the Arctic. What sovereignty is in question? There is no sovereignty in question insofar as the land area is concerned. There hasn't been since 1930. There is no question about the sovereignty of Canada over the continental shelf.[10] There is only a question about the delimitation between itself and its neighbours. The only sovereignty question—I'm not saying it is not important—but the only sovereignty question there is relates to the waters of the Canadian Arctic Archipelago. That is the reason why in 1985, after the crossing of the *Polar Sea*,[11] we were obliged to take a certain legislative measure. This was the enclosure of the Canadian Arctic Archipelago by straight baselines.[12]

If you look at an ordinary Mercator projection or even the much better conformal projection (see map 2 of chapter 12), you get a distorted view. They show the Canadian Arctic Archipelago pointing northward. It does not do so at all. I am so very grateful to the American National Geographic Society, which last October came to the rescue of Canada, for justifying Canada's straight baselines from the legal point of view. Finally, for the first time, we have a world map that shows the Arctic regions, both on the Soviet side and on the Canadian side, pretty well as they are in reality with as little distortion as is possible.

The conformal projection shows the Canadian Arctic Archipelago lying not along the general direction of the Canadian coast, as it should be for the legal justification, but pointing upwards. Whereas the new National Geographic map shows the Archipelago following the general direction from west to east of the northern coast of Canada. Therefore, from the legal point of view, in accordance with the Anglo-Norwegian

Fisheries case of the International Court of Justice of 1951[13] and the Law of the Sea Convention of 1982, Canada, because of the geography, can legally draw those straight baselines as it did. Outside those baselines is the territorial sea of twelve miles and then the two hundred mile exclusive economic zone.

Now the result is this: within the baselines that enclose the Archipelago, the waters, including those of the Northwest Passage, are internal waters of Canada.[14] This means there is no right of passage. This doesn't mean of course that Canada is not going to allow passage. That's not the point. But Canada wants, and I believe rightly so, to keep full and effective control over those waters. It is as simple as that.[15]

I hope you do not think I am unduly nationalistic. But I do believe, sitting as we do between the two superpowers, that we indeed have a right to know what goes on in the waters that we claim to be our own, and I do not think we can depend on our neighbours to tell us that.

With respect to the Northwest Passage, there are three main points of entry, one in the east and two in the west. There is no right of passage through those points of entry. If Canada gives permission, then of course ships may pass through, and that is as it should be.

In spite of the fact that I might appear to be very nationalistic, I firmly believe that we should cooperate with all our neighbours in the Arctic. I taught international law and, in particular, a course on the United Nations, for nearly thirty years, and I have always told my students that international cooperation is indeed the order of the day. I sincerely believe it is. Not only should we cooperate with the United States but we should cooperate with the Soviet Union as well, and also with Denmark, with Norway and with the other Arctic states.

The desirability of cooperation was mentioned this morning. It is nothing new. What was not mentioned is that there is, in my opinion, a legal obligation in Article 122 of the Law of the Sea Convention 1982 for states to cooperate in three respects when there is a regional sea. A regional sea is a semi-enclosed sea, and surely the Arctic Ocean qualifies as a semi-enclosed sea.

The signatories of the convention are to cooperate in the exploitation of the living resources of the regional sea, in the preservation of the marine environment and also in the coordination of their scientific research.[16] Of course, the convention is not, strictly speaking, in force, because it has not yet received the number of necessary ratifications, but it has been signed by Canada and all the Arctic states except the United States, and it does impose, to a certain extent, a legal obligation. Therefore, I do believe that not only should we cooperate because it is in our best interest to cooperate, but we should do so as well because we have a legal obligation.

I conclude by saying a word about the idea of having a nuclear-free zone in the Arctic. I am not sure that we can immediately aim as high as a completely peaceful zone, or, more properly, a so-called zone of peace, but surely we can aim as a first step toward a nuclear-free zone.[17]

This would not be the first nuclear-free zone. Similar nuclear-free zones already exist in the Antarctic, Latin America and the South Pacific, so that if we were to agree in the Arctic, this would be the fourth nuclear-free zone in the world. We have to be careful, of course, before we engage ourselves in that, as it has to be a verifiable nuclear-free zone; but I don't believe this is beyond our imagination. If, after all, the choice in the Arctic is between a sanctuary for submarines and a nuclear-free zone, I would rather have the latter than the former.

Notes

1. These norms are clear enough, but their implementation in a divided world has proven to be most difficult and sometimes impossible. Fortunately, the Arctic has been exempt so far from any real breach of international peace and security. However, there is a latent potential for such a breach, with the two superpowers facing each other across the Arctic Ocean and favouring a forward maritime strategy. Such a strategy might well also pose a threat—or at least be perceived as one—to the political independence and sovereignty of the smaller powers around that same ocean.
2. There are only two exceptions to this prohibition: self-defence and military measures decided upon by the Security Council.
3. The principles pertaining to nuclear weapons are found primarily in two areas of international law: the law of armed conflict, which is largely included in treaty law known as The Hague and Geneva conventions, and special treaties on nuclear weapons.
4. There are close to twenty treaties on arms control, some of which deal specifically with nuclear weapons. The Nonproliferation Treaty is the most important one.
5. The non-nuclear-weapons states also agreed not to manufacture or otherwise acquire nuclear weapons or nuclear explosive devices and to accept safeguards negotiated with the International Atomic Energy Agency to prevent the diversion of nuclear energy from peaceful uses to nuclear weapons. The nuclear-weapons states agreed not to transfer nuclear weapons to any state and agreed to negotiate effective measures relating to the cessation of the nuclear arms race at an early date and to nuclear disarmament.
6. The agreements between the non-nuclear-weapons states and the Nonproliferation Treaty are complementary and represent the legal regime intended to ensure the nonproliferation of nuclear weapons.
7. The Nonproliferation Treaty prohibits nuclear weapons and nuclear explosive devices. It also provides for safeguards in connection with the use of fissionable materials for peaceful purposes. It does not cover nonexplosive military uses. The agreement with the International Atomic Energy Agency permits Canada to obtain exemption from safeguards of nuclear material for nonpeaceful and nonproscribed military activity.
8. The Canadian federal budget delivered to parliament on April 24 and 25, 1989, excluded the acquisition of nuclear-powered submarines. (Editor)
9. Sovereignty is the totality of the various types of jurisdiction possessed by a state within its territorial boundaries, subject only to certain limitations imposed by international law. It extends from the skies to the depths.
10. A state has exclusive jurisdiction over the natural resources of the submarine areas

constituting the natural prolongation of the land territory.

11. The *Polar Sea* is the American coastguard ship that refused to ask permission of Canada to cross the Northwest Passage because the United States considers those waters to be international waters, although Canada claimed them to be internal waters historically belonging to Canada.

12. The enclosure has the legal effect of making the waters enclosed by the baselines internal waters of Canada with no right of innocent passage under customary international law.

13. International Court of Justice Reports (1951): 116.

14. The United States has refused to recognize the baselines and a few other countries have reserved their positions. Their main objection is that the Archipelago does not constitute a coastal archipelago or a fringe of islands and therefore does not meet the geographical requirements of the Anglo-Norwegian Fisheries case, but this is based upon an unduly restrictive interpretation and is supported by a reference to the Mercator or similar projection, which distorts northern latitudes.

15. Canada's national interests in the region are basically three in number: the fragile marine environment, the Inuit population dependent on that environment and the country's national security. If Canada does not protect those interests, all ships, including commercial ships and warships, will have freedom of navigation, and submarines may transit either in their normal mode or submerged.

16. Interestingly, all three areas of cooperation were mentioned by Mr. Gorbachev in his Murmansk speech of October 1, 1987.

17. At least three main points would have to be agreed upon: a complete ban of nuclear weapons, the delimitation of the denuclearized zone (which should include the Kola Peninsula) and the inclusion of all circumpolar states.

Willy Østreng

CHAPTER 11
The Militarization and Security Concept of the Arctic[1]

WILLY ØSTRENG

I would first like to express my gratitude for being invited to this important and timely inquiry. I hope that the example set by the organizers of this event will be picked up by others so that a sensible public debate will bring about the best development of events in the Arctic and in the circumpolar countries.

The purpose of my presentation this afternoon is to discuss how the military utilization of the North has influenced the preconditions for pan-Arctic cooperation in non-military issues in the postwar period. The aim is to identify the dominant features of this relationship and to assess how they might be altered on the basis of changing political circumstances. In order to do this, we need to answer two intertwined questions. The first and foremost question is: Which factors contributed originally to the choice of the Arctic as an area for strategic deployment? The second is: In what way have relations between military and non-military issues been influenced by the security thinking stemming from this choice?

Willy Østreng is director of the Fridtjof Nansen Institute, Norway. He was educated at the Military Academy for the Artillery, Teacher Training College and the University of Oslo. He has been, successively: officer in the Norwegian Army; teacher; research associate; research fellow (Harvard); lecturer, mentor and examiner (Oslo); and research associate (Institute of International Studies, University of California). He has been a member of various professional commissions of trust and the author of numerous books and other publications. He specializes in international relations with emphasis on polar affairs, international security, ocean law, and politics and resource management. He received several notable scholarships between 1968 and 1987.

I turn to the first question first: What factors are behind the militarization of the Arctic? The militarization of the Arctic has been on the increase, in both strategic and geographical terms, over a period of time. Prior to the Second World War, the region was a military vacuum of no strategic utility to anybody. In the 1950s and 1960s the airspace over the Arctic Ocean began to be utilized for strategic deterrence, and today Arctic airspace, ocean areas and indeed outer space are used for this purpose. The interplay between three factors contributes to an explanation of this development:

1. The East-West conflict, which created the political framework for superpower tension and bloc formation
2. Developments in military technology, which produced the atom bomb and other nuclear weapons, as well as long-range means of delivery
3. Geo-strategic factors, which served to indicate the Arctic as a suitable deployment area for strategic high-tech weapons systems

The first two factors were essential conditions for the Arctic to be considered as a deployment area at all, whereas geo-strategic factors explain why the choice was to fall precisely on this region. For that reason I shall focus my attention on the geo-strategic conditions here.

A distinction can be drawn between two main types of geo-strategic factors: universal ones, which place the same geographical constraints on the choice of action of several states in a given region; and state-specific factors, which place more particular geographical constraints on the choice of action of individual states in the same region.

The universal features of the Arctic should be sufficiently well known: the shortest distance between three continents, Asia, Europe and North America, is over the Arctic Ocean (see the small arrows in map 1). None of the major industrial areas in Europe, the Soviet Union, North America or Japan lies more than seven thousand kilometres from the North Pole. Or expressed in another way: eighty percent of the world's industrial production takes place north of 30° north, while seventy percent of the world's major cities (population over one million) are located north of 23½° north. Equally important is the fact that the superpowers can be said to have almost a common border in this area, separated only by the narrow gap of the Bering Strait. In the 1950s it was these geographical facts that indicated the Arctic as a natural route for any nuclear attack using strategic bombers and intercontinental missiles.[2]

The state-specific factors, however, are solely a reflection of Soviet aspirations to be a naval power. As a partly landlocked country, the Soviet Union suffers from clear geographical restrictions in its access to the sea. All of its fleets—the Baltic, the Northern, the Black Sea and, in part, the Pacific—are dependent on passing through straits or narrow sea areas in order to proceed from home territory into high seas and vice versa (see the four large arrows numbered 1 in map 1). Common to all these straits is that they are controlled by powers or constellations with a tradition of tense relations with the Soviet Union. The Kremlin has historical experience to show how several of these straits have been closed for passage in wartime.

The sole exception here is the passage between Svalbard and northern Norway, which, even during the last world war, was not blocked for any appreciable period despite major German efforts. Geographically speaking, this passage is considerably broader than the others, being 345 nautical miles in breadth. In terms of climatic conditions, it has the advantage of remaining ice-free throughout the year. And in politico-military terms, it appears particularly attractive because its northern limit, the Svalbard Archipelago, is partly demilitarized, while its southern shore, mainland Norway, is subject to self-imposed military restrictions (for example, as to base and nuclear policy).[3]

Against this background then, the Northern Fleet, based at the Kola Peninsula, was gradually developed to become the most modern of Soviet fleets. Today it ranks second to none with regard to survivability and to strategic retaliatory capability. At its disposal are sixty-six percent of the Soviet Union's strategic submarines, sixty-seven percent of her submarine-launched ballistic missiles, seventy-six percent of her warheads and seventy-three percent of the megatonnage available to all of the fleets.[4] However, the country's general geographical handicap was to catch up with this fleet as well.

In the late 1960s and early 1970s sea-based deterrence in the Northern Fleet was taken care of by Soviet Yankee submarines equipped with short-range missiles. To reach targets in the United States, these vessels would have had to sail through the GIUK gap—between Greenland, Iceland and the United Kingdom—and then assume launch positions off the east coast of the United States (see large arrow numbered 2 in map 1). The problem here, as seen from a Soviet point of view, was that the United States practised a barrier strategy in this area so as to be able to intercept Soviet submarines in transit in time of war. It was vital for the Soviet Union to counteract this threat since it concerned the Soviet Union's own retaliatory capability. The response came in 1972 with the Soviet deployment of the first Delta-class submarine, equipped with SSN-8 missiles with a range of forty-three hundred nautical miles. From then on, Delta submarines were capable of striking any target in Europe, North America or even China from launch positions in Arctic waters. By stationing its Delta—and later Typhoon—submarines north of the GIUK gap and under the polar ice, the Soviet Union avoided countermeasures that could cripple the strategic combat potential.[5]

The withdrawal of the Northern Fleet's strategic forces to Arctic

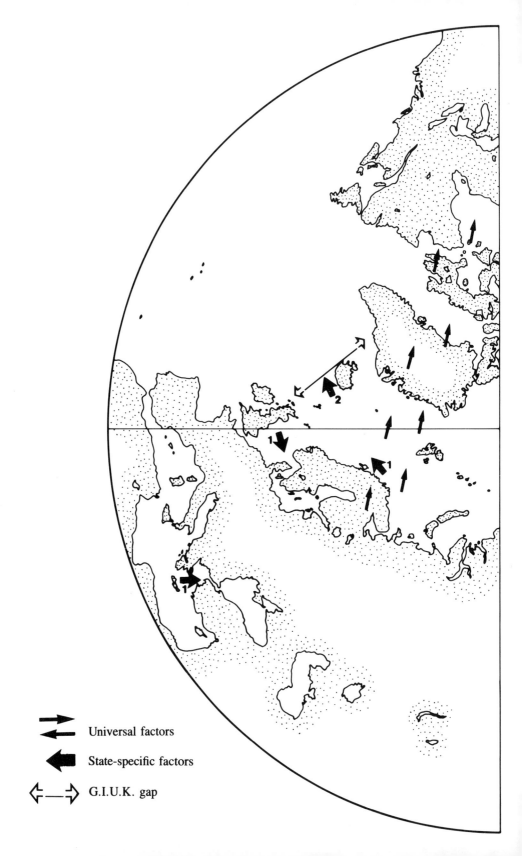

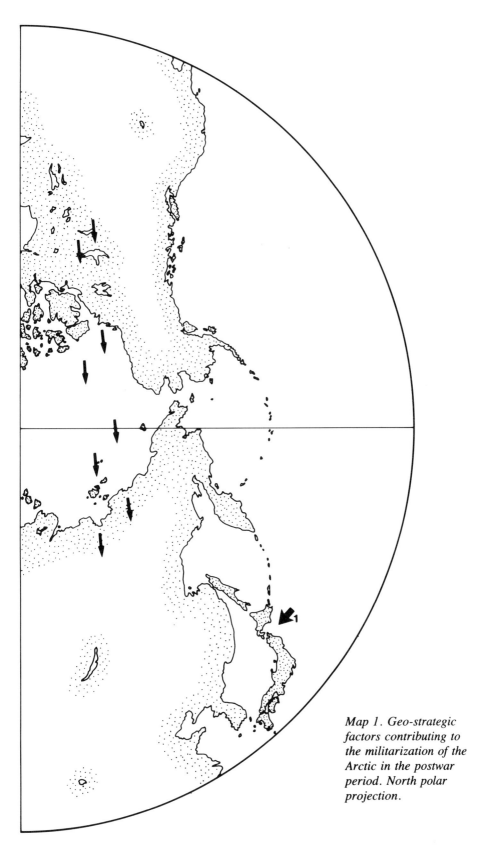

Map 1. Geo-strategic factors contributing to the militarization of the Arctic in the postwar period. North polar projection.

waters—the rear deployment strategy—has led the United States in recent years to follow suit, introducing military countermeasures in the same waters. In this way the Arctic has gradually been transformed from a military vacuum prior to the Second World War to a military flank in the 1950–1970 period and to a military front in the 1980s.[6]

Geographical constraints have thus, given the prevailing political and military technological circumstances, predetermined the Arctic for militarization. As a contra-factual hypothesis, it may be maintained that the superpowers had no alternative but to put the region's geographical advantages to military use. Viewed in this light, it seems reasonable to conclude that any change in the situation in the Arctic can come about only as a result of fundamental changes in East-West relations, which long precluded the possibility of non-military cooperation.

I will now turn to the second question, the interrelationship between militarization and non-military issues in the Arctic.

Before 1970 science was the only non-military area to be termed international in the sense that several states conducted research and collectively had an interest in each other's activities. Among researchers, there was a widespread opinion that "science and scientists have a kind of objectivity which is congenial to cooperation. The nature of science is not only conducive to cooperation, but, indeed, demands it, for no man, no nation, has a monopoly of science."[7] Thus, science provided a conditional point of departure for non-military cooperation among states in the Arctic. Let us therefore use science and military strategy as an illustration of the interrelationship between military and civil issues.

In the 1950s the Arctic Ocean became a tension field in power politics—a taboo area where constraints were also placed on scientists. The few foreign scientists who carried out field studies in the far North of Canada during the 1950s and 1960s, for instance, were subjected to security restrictions in their work. Similar restrictions were applied in Alaska and Greenland as well.[8] Furthermore, at an early stage, Soviet authorities prohibited their own scientists from participating in international organizations for Arctic research.

The sole area in the Arctic where researchers have enjoyed freedom of research throughout the entire postwar period is on Svalbard. Here it has been Norwegian policy, in line with the spirit of the Svalbard Treaty of 1920, to offer researchers of all political colours, unimpeded access and equal working opportunities.[9]

Everywhere else research in the Arctic was to develop rapidly in a nationalistic, partly chauvinistic, direction. Research was mainly organized under the auspices of national regimes; it was only rarely binational, and even more rarely multinational. Most scientists on both sides were engaged in applied research to achieve military superiority in the region.[10]

This situation came about not only because of the need for applied knowledge but because, during this period, the East-West conflict was of a decidedly hegemonic nature. By this I mean a conflict taking place between competing political systems that develops cumulatively, in that conflict in one issue automatically spills over into the others. Purely hegemonic conflicts do not permit issues to be kept outside the conflict realm: here military conflict spreads to embrace the entire range of interests and all points of contact between the parties. In times of crisis hegemonic conflicts demand unambiguous national answers to what are interpreted as challenges from the adversary. No sector or segment of society, not even research, can free itself totally from conflicts of this type.[11]

Consequently the hegemonic features of the Cold War contributed at an early stage to the creation of what we could call a fully integrated, multi-dimensional security concept for the Arctic. The linkage between military and scientific fields became almost absolute, with few or no distinctions made between the two areas. This situation remained more or less unchanged until the beginning of the 1980s. As a consequence, scientific cooperation across the East-West divide could hardly be conceived of as a realistic alternative to the ongoing national endeavours.

[Editor's note: Passages in italics were in the written speech but were not delivered orally because of time constraints.]

Political circumstances, however, began to change. With the spirit of détente in the 1970s, more and more indications came that the hegemonic conflict was in the process of weakening in power and intensity. Linkages between military and non-military issues became less pronounced, and some cautious attempts at cooperative ventures were initiated in civilian issues. For example, a cooperative agreement was signed in 1972 between the United States and the Soviet Union on environmental protection, in which the Arctic was included. The policy of détente was a contributory factor here. However, the foundering of détente was to have the reverse effect in the second half of the 1970s.

Two conditions may help to further the cooperative trend introduced by détente. The first of these conditions is change in the Law of the Sea, which, in the wake of the ongoing third United Nations Law of the Sea Conference, produced two hundred nautical mile economic zones for the administration of living and mineral resources. In order to avoid conflict, these zones "forced" adjacent and opposite states to cooperate in the management of joint fish stocks and petroleum structures straddling the delimitation line between them. This applied to the Arctic.[12]

The second condition helping to further the cooperative trend is changes in the concept of Arctic security.

On October 1, 1987, General Secretary Mikhail Gorbachev gave a speech in Murmansk in which he signalled the will to distinguish more sharply between military and non-military issues. The aim, which he sketched out, was to bring about "a radical lowering of the confrontation level in the area. Let the northern part of the globe, the Arctic, become a zone of peace. Let the North Pole become a pole of peace."[13]

Gorbachev spoke, for instance, in favour of cooperation in the exploitation of natural resources and for the establishment of a unified energy program for the northern areas. Furthermore, he invited Norway and Canada to participate in joint ventures and projects for the exploitation of oil and gas on the northern Soviet continental shelf. Other states too were invited to negotiate such cooperative ventures. Moreover, emphasis was placed on cooperation on environmental issues, and the Northern Sea Route was, under certain conditions, to be placed at the disposal of foreign ships, which were offered the assistance of Soviet icebreakers.

Also with respect to research, a break with the former isolationist policy of the Soviet Union was signalled. First of all, the importance of international cooperation was stressed, as was the need for the exchange of scientific experience from the Arctic. Furthermore, Gorbachev spoke in favour of adjacent states coming together at an international conference for the coordination of future research and for discussion of the establishment of an Arctic research council. This conference took place in Leningrad in December 1988. The departure from the previous period is clear: the line of "national self-sufficiency" is to be supplemented by international cooperation and coordination; national organization is to be supplemented by an international research council; secrecy is to be reduced through the exchange of Arctic research experience.

On all these points, Gorbachev has actually decoupled non-military issues from the integrated concept of security that dominated the period from 1945 to 1980. And there is much to indicate that Gorbachev was in earnest concerning these proposals.

To the amazement of many observers, the Soviets gave their approval to the 1986 SCAR[14] initiative to elucidate the possibilities for future scientific cooperation in the North. The report prepared in this connection speaks in favour of the establishment of a nongovernmental international Arctic science committee to "promote international cooperation in scientific research in arctic areas." It is further proposed that there be established an intergovernmental forum on Arctic science issues, where "representatives of the governments... should discuss the feasibility of establishing a system of regular, structured discussions and liaison on arctic science matters."[15] During the Leningrad meeting, the Planning Group of the International Arctic Science Committee (IASC) agreed on a new proposal, entitled "Founding Articles for an International Arctic Science Committee," to be endorsed by representatives of the founding countries, the Soviet Union included.

It would appear that the Soviet Union has also modified its policy of denying Western researchers access to Soviet Arctic soil. In the Canada–U.S.S.R. Arctic Science Exchange Program of 1984 and in the 1988 Norwegian-Soviet Agreement Concerning Technical-Scientific Cooperation on Exploration of the Arctic and the Northern Areas, provision is made for the establishment of cooperation that is to be balanced in terms of scholarly issues and geography, and for scientists of both parties to take part in field studies on each other's territory.

These changes are especially interesting as they have come at a time when the Soviet Union would in fact have more to protect—or so one might suppose—against prying Western eyes than ever before, not least in connection with the rear deployment of Soviet strategic submarines (SSBNs).[16] *In the above-mentioned agreement with Norway, the Soviet State Committee for Science and Technology accepted that joint oceanographic studies are to be carried out in, among others, the waters between Greenland and Svalbard.*

This strait, the Fram Strait, is one of the main sea routes for Soviet strategic submarines travelling northward to the Arctic Ocean and southward to the Norwegian Sea. Oceanographic research cooperation across the East-West divide in this area can therefore be seen as a military risk to the Soviets. On the other hand, there is also a military advantage in getting to know what your opponent knows and thinks in areas of strategic relevance to both sides. Seen in this perspective, cooperation becomes a means of obtaining access to the opponent's knowledge and insight—if you can't beat him, join him. Another explanation might be that the Soviet Union is consciously expanding the range of themes for research cooperation in order to achieve greater credibility for its rapprochement policies. Thus, cooperation is considered acceptable because the political gains are seen to outweigh the military disadvantages. No matter what the explanation may be, we can note that the Soviet Union has taken a step towards enlarging the framework for scientific cooperation with Western countries. This shows that political will and changes in political circumstances can, in themselves, provide the preconditions for international cooperation even in areas formerly reserved for applied research under national auspices. However, there will always be a limit to how far the authorities can go in accepting research themes that are to be made the subject of international cooperation. The need will still remain for privileged information meant solely for national use. What politicians can do is to adjust these limits, either in a more liberal or in a more restrictive direction. What now appears to be happening on the Soviet side is an adjustment in a more liberal direction.

All this indicates that the older Marxist concept of "totally integrated security" is breaking up, at least for those issues that, in the strictly military sense, do not concern the military security of the state. If this process continues, the Soviet Union will be left with a decoupled concept of security around a military strategic core. In other words, it seems that the period of meshing military with non-military issues may be coming to an end for the Soviet Union. This is not to say that the Gorbachev regime has repudiated the traditional Marxist concept of comprehensive security. What we are witnessing is more likely a shift of tactics rather than of concepts. By decoupling the two sets of issues, non-military security may be achieved by employing, among other means, the cooperative potential of individual civil issues. Military rivalry no longer seems to be the dominant interest in defining the content and pace of achieving non-military security. By this tactic, comprehensive security may be achieved asymmetrically, in that the level of security may differ between individual issues. Previously, it did not. Whether this development will serve to improve the conditions for East-West cooperation will depend, among other things, upon the content of the Western concept of security.

During the Second World War, the U.S. Vice-President, Henry A. Wallace, proposed that the United States should lead the way in establishing an Arctic Treaty for, among other things, scientific exploration and cooperation among the Arctic states.[17] *However, the Cold War prevented the implementation of this proposal. During the International Geophysical Year of 1957–1958, time-limited cooperation was initiated in the region, but it was not until the early 1970s that the Western states were again to make a serious cooperative initiative towards the Soviet Union. In 1970 for example, the United States invited several nations to participate in the Arctic Ice Dynamics Joint Experiment—an ambitious project for studying thermo-balances and the relationship between ice cover and atmosphere. However, only two countries, Canada and Japan, accepted the U.S. invitation.*[18] *The Soviets refused the invitation on the ground that "the economic and scientific reasons for investigating the Arctic Ocean are intertwined with the military ones, which have elicited great interest on the part of the U.S. Navy."*[19] *Only two years later Norway took the initiative in negotiations for a polar bear research and conservation agreement between the United States, Canada, Denmark, the U.S.S.R. and Norway.*

The implementation of the polar bear agreement signed in 1973 also helps to illustrate the attitudes to cooperation held by the various parties. Among other things, the exchange of information between Western delegates has been profuse and in line with the letter and spirit of the agreement. By contrast, the Soviet contribution has been so negligible as to give rise to the following Western reaction: "Whether only little research has been carried out in the Soviet Union, or whether the data exchange is being hampered by governmental red tape it is difficult to say."[20]

It is also worth noting that representatives from the United States, Canada, Denmark, Sweden, Finland, Iceland and Norway have all supported the proposal to establish an international Arctic science committee. Thus, there is reason to maintain that the Arctic states of the West have remained open for functional cooperation in non-military areas throughout the past twenty years.

The interesting point about these examples is that they show how the Western states in fact started trimming down and decoupling their security concept a good ten years before the Kremlin did. As long as military and non-military issues are combined into an integrated security concept, there is always the risk of militarizing non-military issues—as indeed happened during the years from 1945 to 1980. All the Arctic states now find themselves in a process of moving away from a totally integrated hegemonic conflict to a differentiated situation in which military strategic conflict may come to exist side by side with non-military cooperation.

To conclude, the paradox of the present situation is that the prospects of establishing pan-Arctic cooperation in non-military issues seem to have improved at a time when the militarization of the region is on a steep increase. Ten years ago the opposite would have been the case. This implies that changes in political circumstances, under certain conditions, may provide the necessary impetus for international cooperation to emerge, even in issues formally reserved exclusively for national consumption and guarded by measures of secrecy.

Notes

1. This paper draws heavily on another paper of mine; see Willy Østreng, "Political-Military Relations among the Ice States: The Conceptional Basis of State Behaviour," presented at the International Conference on Arctic Cooperation, Toronto, October 26–28, 1988, co-sponsored by the Canadian Institute for International Peace and Security and Science for Peace. It also resembles a speech I made during the fall meeting of the American Geophysical Union, San Francisco, December 5–9, 1988.

2. Only forward bases on the territory of other states could bring the opponent's territory within reach of one's own weapons. The Soviet Union lacked such bases and thus had no alternative to the Arctic. The United States, on the other hand, had the option of using both forward bases and Arctic airspace. However, seen from the U.S. perspective, it was important to make use of the Arctic because bases on the territory of other states, depending on the political climate, might have been wound up at any stage. Polar attack routes were also preferable for Intercontinental Ballistic Missiles (ICBMs), since the shorter the distance, the greater the accuracy. In this way, geography laid close to equal constraints on both sides in their choice of strategy.

3. Willy Østreng, *Politics in High Latitudes: The Svalbard Archipelago* (Montreal: McGill–Queen's University Press, 1978): 44–60.

4. Thomas Ries, "The Soviet Operation Command Structure and its Application to Fenno Scandia," Report NUPI (Oslo: Norsk Utenrikspolitisk Institut, August 20, 1986): 61–62.

5. Willy Østreng: "The Soviet Union in Arctic Waters," Occasional Paper No. 36, The

Law of the Sea Institute (Honolulu, 1987), chap. 3, pp. 27–41.

6. Ibid., pp. 43–45. Military necessity and geographical constraints were thus to indicate the Arctic as a deployment area for strategic bombers and ICBMs, and this at an early stage. Here the universal geo-strategic factors were equally relevant for both the United States and the Soviet Union. It was, however, the Soviet Union's state-specific problems as a naval power that led to the diversification in militarization of the region—first, its need to raise the status and importance of the Northern Fleet in relation to the other Soviet fleets; and, second, its need to use Arctic waters as a deployment area for strategic submarines. Here political and military choices have had to be adjusted to an unchanging geographical reality.

7. Hugh Odishaw, "International Cooperation," in *Science and Technology*, unnumbered, undated, in the archives of FNI, p. 28.

8. Trevor Lloyd, "International Cooperation in Arctic Science and Disarmament," mimeographed article in the library of the Canadian Institute of International Affairs, Toronto, March 1969. Illustrative of this situation is the description provided by Professor Trevor Lloyd: "It must be acknowledged that there has been a definite barrier between the scientists of the Soviet Union, and associated countries and other countries. There has been no easy interchange between 'western' group members and the rest.... Although Polish and Czechoslovak scientists have worked with Soviet scientists in Antarctica there is no record of this being done in the North, whether in the Soviet Arctic or elsewhere. The Soviet Arctic has not yet been opened to western scientists, although a few individuals have made short visits there."

9. Willy Østreng (1978), op. cit., pp. 44–60.

10. Such a trend reduced the possibilities of establishing cooperation across the East-West divide. No one wished to share with the enemy research results that could have been put to counter-use in case of war. Moreover, as a starting point, it was not desirable to reveal to the opponent how much one knew—not to mention revealing one's own weaknesses. Likewise, each side was wary of supplying the opponent with information that could have promoted his interests in the region. Politicians viewed any possible cooperation as something that could well have yielded such undesirable results. Thus, the desires of researchers for scientific cooperation ran aground on the reef of politics. See Willy Østreng, "Polar Science and Politics: Close Twins or Opposite Poles in International Cooperation?" Paper presented at the international symposium, The Management of International Resources: Scientific Input and the Role of Scientific Cooperation. Fridtjof Nansen Institute, October 10–11, 1988.

11. The requirement of loyalty is absolute. The state has control of the sectors and formulates whatever answers to the conflict it considers necessary. These responses bind the actions and behaviour of those actors whose own interests and priorities favour cooperation across national borders.

12. See Willy Østreng, "Delimitation arrangements in Arctic Seas. Cases of precedence or securing of strategic/economic interests?" *Marine Policy* (April 1986): 132–154.

13. Speech by Mikhail Gorbachev in Murmansk, October 1, 1987. Appendix to *Sovjetnytt*, no. 26 (1987): 11. The "Murmansk Program," consisting of six items, included both military and non-military issues.

14. The Scientific Committee on Antarctic Research, a committee of the International Council of Scientific Unions.

15. E. F. Roots, O. Rogne and J. Taagholt, "International Communication and Coordination in Arctic Science: A Proposal for Action," (Oslo, November 17, 1987): 2.

16. See Willy Østreng, "The Barents Sea in Soviet Rear-deployment Strategy," *Naval Forces* (1988/89). Forthcoming.

17. Herman Pollack and Peter Anderson, "United States Policy for the Arctic," *Arctic Bulletin*, Vol. 1, No. 3 (1973).

18. *Aidjex Bulletin,* No. 15 (Seattle: Aidjex Scientific Plan, August 1972).
19. A. F. Treshnikov, E. Borisenkov, N. A. Volkov and E. G. Nikiforov, "The American Arctic Ice Dynamics Joint Experiment Project," A. F. Treshnikov (ed.), *Problems of the Arctic and Antarctic,* Vol. 37 (Leningrad: 1971).
20. Tore Gjelsvik, "Science and Politics in Polar Areas," mimeographed article in the archives of the FNI, October 1985.

Arctic "Militarization": The Canadian Dilemma

Keith R. Greenaway

CHAPTER 12
Arctic "Militarization": The Canadian Dilemma

KEITH R. GREENAWAY

I am most thankful that Professor Pharand came before me because to some I might sound like a rabid nationalist, but with Donat leading the way I feel more comfortable about that. Also, in case some of the bureaucrats in Ottawa and some of the politicians may be a bit nervous about what I might be saying and feel that I should be thrown out of the country for insubordination, I am going to watch where Stephen Lewis heads and, if you see both of us getting on the same aircraft or the same boat, you will know what has happened!

Brigadier-General Keith R. Greenaway (retired) is a private citizen and consultant on northern matters in Ottawa. He joined the Royal Canadian Air Force in May 1940 and served on instructional duties throughout the Second World War. In 1945 he began an innovative and experimental career phase with an assignment with the U.S. Navy on pressure pattern flights over the North Atlantic. For the next fifteen years he worked, sometimes in association with U.S. agencies, on problems of high-latitude and long-distance navigation. He commanded the RCAF Central Navigation School and RCAF Station Clinton. On retirement from the Canadian Forces, he began an association with the Advisory Committee on Northern Development and the Department of Indian Affairs and Northern Development. Later he took a leading role in private ventures in remote sensing and resource surveying. He has received numerous honours in the United States and Malaysia (where he served on assignment to External Affairs) as well as in Canada. These include the McKee Trophy, the Massey Medal, the Order of Icarus, the Order of Canada, an honorary doctorate, and membership in Canada's Aviation Hall of Fame.

I'm going to turn to a domestic situation that certainly has international overtones, but one that is not well known. Within that context, a short time ago, Professor Cox of Queen's University very succinctly summed up the difficult choices confronting Canada in the circumpolar region by stating that "the challenge facing Canada is to devise policies which protect Canadian sovereignty, respond to legitimate U.S. security needs, and make a contribution to the stability of the superpower relationship in the circumpolar Arctic." Quite obviously there are many aspects of this challenging task that relate to what is often perceived to be the militarization of the Arctic.

The subject I wish to dwell on today is the surveillance of airspace, where concern over sovereignty has not grasped the attention of the public or the politicians to the same degree as that associated with the maritime environment. However, surveillance of our airspace is no less vital to preserving the sovereign state of Canada. The subject is much too large and complex to review in detail today. Therefore, in the time allotted, I wish to focus on one of the most glaring weaknesses in surveillance and control of Canadian airspace for the purpose of safeguarding our sovereignty and also to attempt to indicate where the problem lies.

The civil air control service for the Canadian sector of the Arctic Basin is divided into two sections, one encompassing the area south of latitude 72° N and the other encompassing the area north of latitude 72° N to the North Pole for flights above about nine thousand metres (map 2).

The introduction of the civil air control service through the International Civil Aviation Organization in 1970 greatly enhanced Canadian sovereignty over its polar frontier. The system, however, is designed to handle cooperative aircraft only. In other words, it controls aircraft for which flight plans have been filed and whose pilots report flight progress.

Radar surveillance by the North American Aerospace Defence Command's (NORAD) new North Warning System does not extend beyond about 72° N, which is the southern boundary of the northern section (map 3). Hence there is no effective sovereignty control of the airspace in the vast strategic region of the Canadian Arctic Archipelago, which is about half a million square kilometres, or a little smaller than the province of Alberta. We know that about six thousand aircraft pass through this airspace annually. How many others use it, we don't know, nor do we know how many cooperative aircraft positions are reported incorrectly. In fact I've often heard it said that this is the only area in the world where every aircraft is on track and on time.

Why do we have this major weakness, as well as others, in the surveillance and control of our airspace, when the two responsible federal agencies are spending a total of about $2 billion over the next few years on upgrading surveillance and control services? The Department of National Defence argues that the location of the North Warning System is quite satisfactory, as it meets NORAD's requirements and therefore satisfies Canadian needs. The Department of Transport is preoccupied with upgrading facilities in southern Canada and with taking over radar surveillance in regions previously serviced by the Department of National Defence through the Pinetree Line.

The problem of not having a national perspective and a long-term view of surveillance and control of the airspace appears to be with the implementation of our Aeronautics Act. Under the act, responsibility for the protection of Canadian sovereignty, as it applies to the surveillance of Canadian airspace, is vested in both the Minister of Transport and the Minister of National Defence. The Minister of Transport is responsible for the control of air traffic in Canadian airspace except in matters relating to defence, in which case the responsibility is that of the Minister of Defence. Hence, the Department of Transport's interest lies in the control of cooperative air traffic, whereas the interests of the Department of National Defence lie in the detection and control of noncooperative air traffic.

The interests of the Department of National Defence are shared with the United States through NORAD—a forum in which Canadian sovereignty is usually viewed more or less as a nuisance—and is expressed almost solely in terms of protecting the deterrent forces of the United States.

In practice, the shared responsibility for the surveillance and control of Canada's airspace has been exercised quite separately by the Department of Transport and the Department of National Defence. Regrettably, this has led to each agency developing its own priorities, programs, organizations and systems, as well as equipment. Thus, in the process, each agency appears to have emphasized differences rather than highlighting important similarities; as a sad result, the manifestation of Canada's sovereignty over its airspace has been compromised.

In 1985 Canada signed the North American Air Defence Modernization Agreement with the United States. This agreement is basically a technical update of the original Distant Early Warning (DEW) Line, with additional radars on Baffin Island and along the Labrador coast, and with new control features. I might mention that, until these radars were installed, a large gap existed along the southern Baffin and Labrador coasts. Thousands of aircraft have come through this gap without verification until they were well inside our country. In many cases it would be several hours before they were verified by the Pinetree Line further south. This was not realized by the public.

Periodically, and in times of tension, the U.S. Air Force Airborne

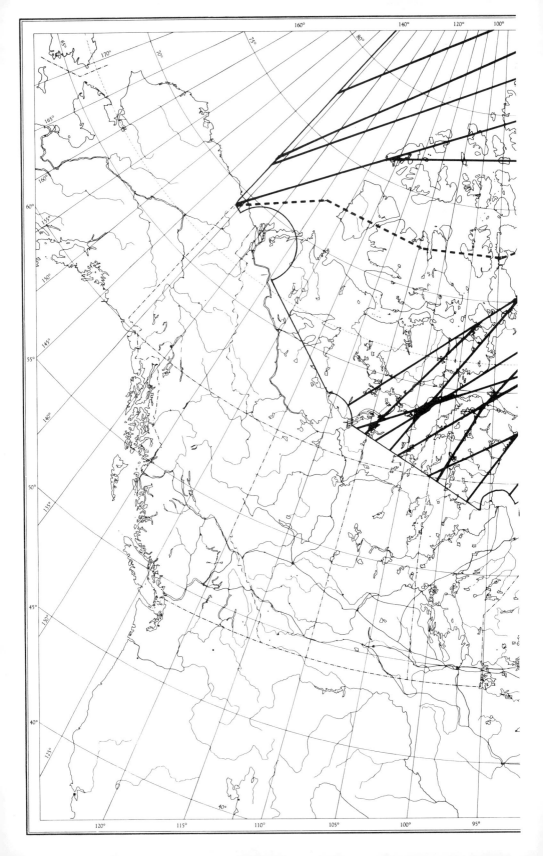

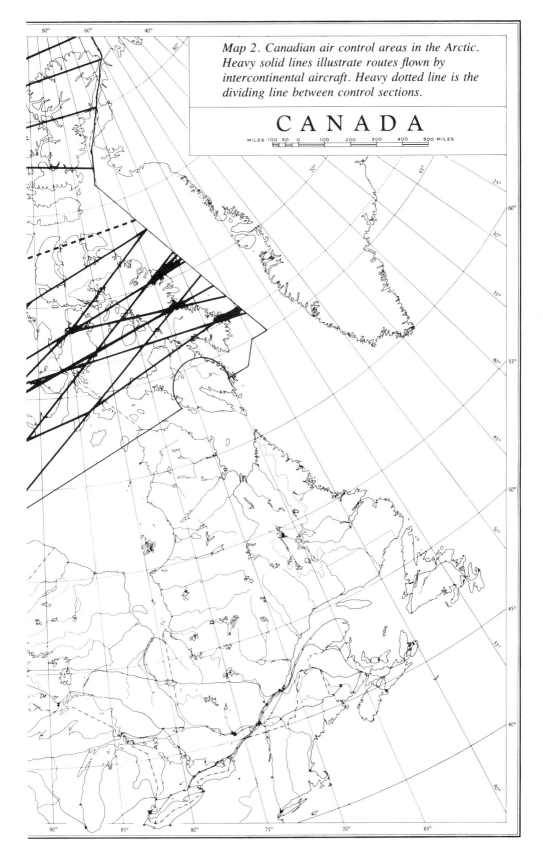

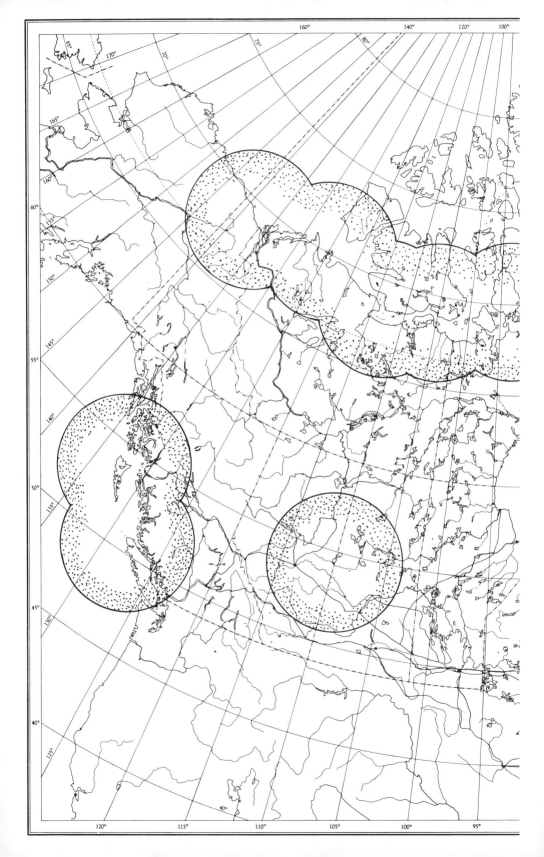

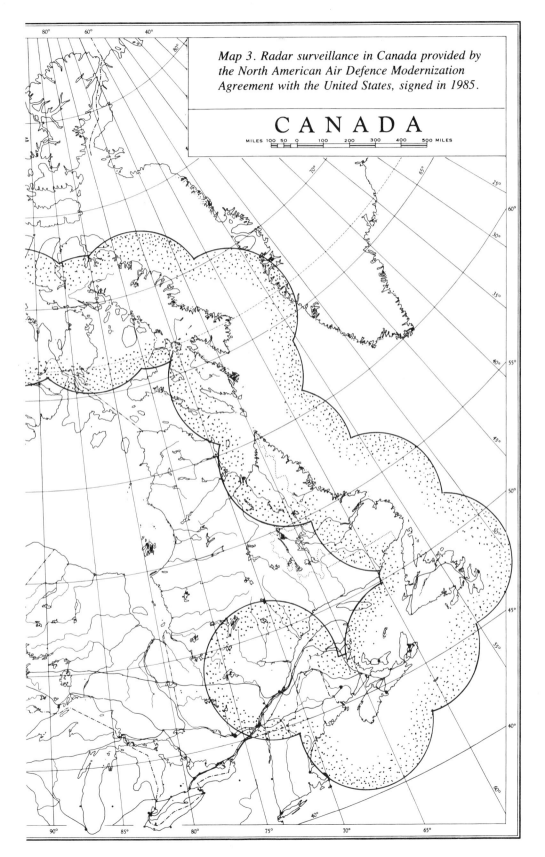
Map 3. Radar surveillance in Canada provided by the North American Air Defence Modernization Agreement with the United States, signed in 1985.

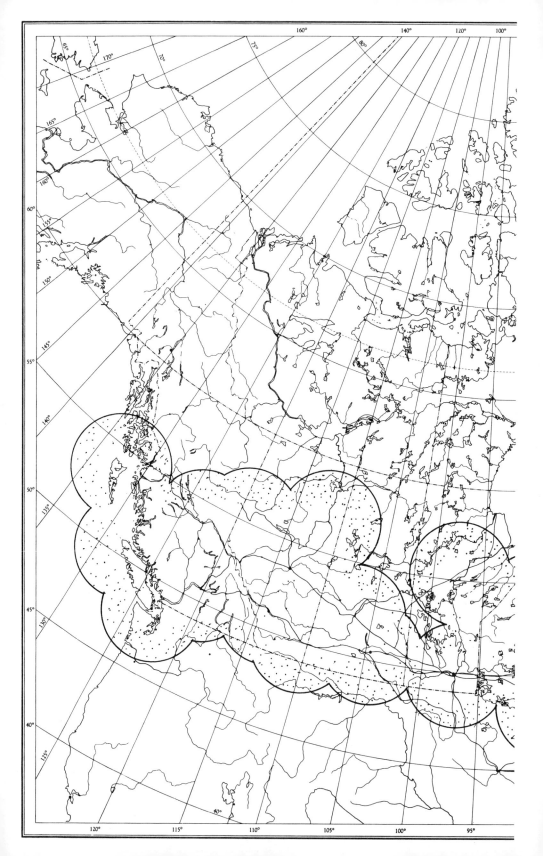

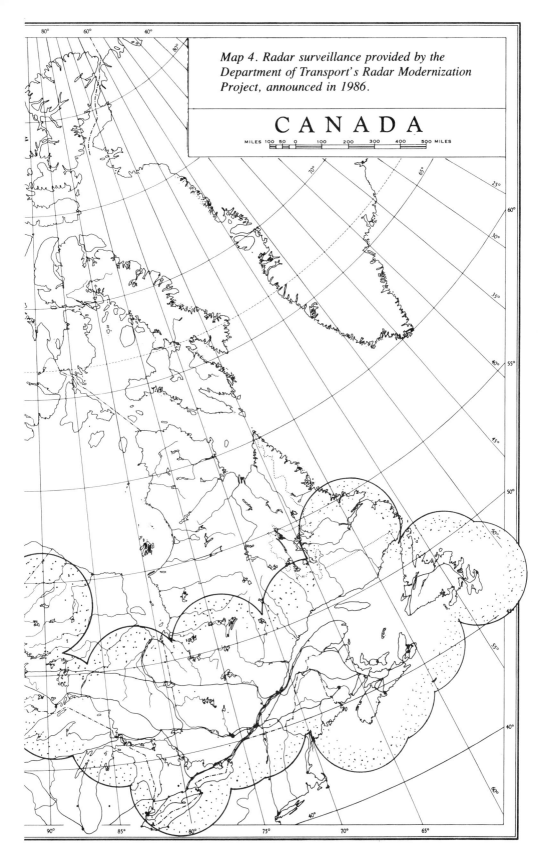

Map 4. Radar surveillance provided by the Department of Transport's Radar Modernization Project, announced in 1986.

CANADA

Warning Control aircraft assigned to NORAD will supplement the surveillance provided by the ground-based system under the 1985 agreement. This supplementary surveillance is planned for areas in front of and behind the North Warning System. The Pinetree Line in southern Canada has been closed down, and with the establishment of "Forward Operating Locations" as far north as Inuvik and Iqaluit, interception of intruders will take place farther north and not over the more densely populated areas of southern Canada as was the case for decades, apparently without any opposition from our politicians or the public.

The Canadian financial contribution to the modernization program is about $600 million. In addition, Canada is assuming full responsibility for the Canadian sector of the North Warning System, including operating costs for all facilities in Canada.

This new agreement is a vast improvement over the previous arrangements, in that gaping holes in surveillance along the northeast flank have been plugged and Canada has assumed more responsibility for that part of the NORAD system located in Canada. On the other hand, it fails to enhance sovereignty over the airspace in Arctic Canada, a region that has been undergoing major social and economic changes and has increasing strategic significance. One cannot take for granted that just by being a responsible partner in NORAD the Department of National Defence is fulfilling its obligations to protect the sovereignty of Canada in Canada's frontier airspace, as prescribed by the Aeronautics Act.

The surveillance provided by the Department of Transport's Radar Modernization Project, announced in 1986 and to be completed in 1992 at a cost of about $1 billion, is part of the Canadian Airspace System Plan, the implementation of which extends into the year 2000 at a cost of about $5 billion (map 4). However, it has no provision for improving radar surveillance on the polar frontier beyond that provided by the NORAD North Warning System. It is designed to cover only the provinces, particularly their southern portions, and nowhere extends above 60° N.

Redundancy in radar coverage across parts of southern Canada, which had existed in the past, was eliminated with the shutting down of the military Pinetree radars in 1988. In some cases, gaps in coverage were created with the shutdown of Pinetree radars, but these are now being filled under the Department of Transport's Radar Modernization Project. Intercept capability across southern Canada has been eliminated, and reliance has been placed on the U.S. Joint Surveillance System for control of unidentified aircraft crossing the border in either direction.

To summarize, we have a civilian control system covering all Canadian airspace but two radar surveillance systems: a civil one confined to southern Canada and a joint Canada–U.S. military one (NORAD) that is farther north but that fails to provide coverage on the Arctic periphery of the country.

In the United States, a joint military and civil surveillance system provides coverage around the periphery of the country. Why not a joint system for Canada? Although the Department of National Defence and the Department of Transport may apply the data differently, much of it is derived, processed, transmitted and displayed by identical or very similar equipment and technology.

The plea for a joint civil-military system in Canada is not new. About a decade ago General Norm Magnusson, a Canadian airman with considerable experience in this field, stated in the *Defence Quarterly* that a national system would serve national aims by:

• reducing the number of radars in Canada to a network of selected Department of Transport and Department of National Defence sites for en route tracking of aircraft by the Department of Transport and surveillance of sovereign airspace by the Department of National Defence;

• establishing a common communications network complementary to the common radar system—thus eliminating present and planned duplicate facilities;

• establishing automatic control centres in Canada jointly operated by the Department of Transport and the Department of National Defence for the control of both civil and military air traffic;

• retaining a viable capability to cooperate with the United States in the defence of the continent.

Although this plea was made when there was considerable duplication of facilities in southern Canada, it is still valid. In fact, if a national system had been put in place, it is most unlikely that the inability to exercise sovereignty over a large section of Canada's strategic airspace would exist today. In a similar vein, a recent report by a working group of the Canadian Institute of International Affairs concluded that the inability of Canada to exercise sovereignty over its northern airspace appears to be the result of lack of appreciation of the national interests involved and of inadequate coordination of measures to protect those interests. The working group recommended that a National Air Surveillance and Control Plan be prepared as a first step towards establishing a national system that would satisfy civil and military airspace surveillance and control needs. Moreover, with the approach of the era of space-based surveillance radars, the need for a national surveillance plan becomes urgent. The integration of Canadian civil and military requirements is a prerequisite to any discussion with the United States on any possible joint surveillance arrangement.

In closing, I wish to leave with you several very cogent thoughts. They were expressed by General Paul Manson, now Chief of the Defence Staff, when referring to Canada's airspace sovereignty in peacetime before the Special Senate Committee on National Defence. General Manson said:

> I instinctively, as I think most Canadians do, have a feeling that unless we have the capability of controlling our airspace, that is of knowing the presence of an intruder and being able to intercept and identify that intruder, to enforce our sovereignty in airspace, there is something lacking in the composition of the Canadian nation.

In turning to the U.S. Joint Surveillance System, General Manson went on to say:

> It is a comprehensive military/civil radar chain around the interior periphery of the United States. I believe their motivation in putting that in place relates precisely to the point that we have just been talking about, that is, that no nation can really declare that it has full control and full jurisdiction over its own airspace unless it has the capability of controlling and identifying air traffic within that zone. The Americans, I know, feel very strongly about this.

Finally, the dilemma in my opinion is whether Canadians hold similar views about the safeguarding of sovereignty in Canada's airspace and, if so, whether they have the will and the political courage to do something about it.[1]

Notes

1. Those who are interested in delving into this important but complex national issue may wish to read the following: "Canada's Territorial Air Defence," Report of the Special Committee of the Senate on National Defence (Ottawa, 1985); An Aurora paper on "Canada and NORAD 1958–1978: A Cautionary Retrospective" by David Cox, Canada Centre for Arms Control and Disarmament, 1985; "Offering Up Canada's North" in *Northern Perspectives,* Canadian Arctic Resources Committee (Sept.–Oct. 1986); "The North and Canada's International Relations," a report by the Canadian Institute of International Affairs, Ottawa Branch (March 1988).

CHAPTER 13
Question Session

KEITH R. GREENAWAY, WILLY ØSTRENG,
DONAT PHARAND, GORDON ROBERTSON
AND WALTER SLIPCHENKO

Linda Hughes
Professor Pharand, you put forward a very strong argument why we legally have sovereignty over the waters of the Arctic Archipelago. Is there a problem asserting that sovereignty if we don't have a single vessel that can patrol those waters year-round and if we don't have surveillance of those waters?

Donat Pharand
I believe the answer has to be in the affirmative, in the sense that our legislation would not be credible for very long. After all, it would be only paper sovereignty. That is not good enough. We have to be in a position to exercise what in international law is called "effective control." Now, mind you, as I have pointed out, this does not mean that all means and methods of exercising effective control are legitimate. Nevertheless we should at least exercise a control that will permit Canada to know whether in the Northwest Passage, or anywhere else for that matter, there are submarines and whether they are friendly or unfriendly and, if unfriendly, to take either collective or unilateral measures. I do not think that we could, strictly speaking, take much by way of unilateral measures in any event. That is the reason why I don't know that it is really necessary for us to have nuclear-powered submarines, because what more are you going to do? If you have sonar mechanisms to detect at the entrances, let us say, of Lancaster Sound in the east and Amundsen Gulf and McClure Strait in the west, and you spot an unwanted submarine, what is the difference? What do you want to do, whether you have detected it by way of a mechanism or by way of a nuclear submarine? If the submarine is nuclear-powered,

Question Session

Adrienne Clarkson

what action are you going to take? In my opinion, you are not going to start shooting at it and declare a third world war. What you then would do, I suggest—I think this is about the only thing you could do—is simply to publicize the matter or bring it to the attention of the Security Council, or both. That is the route in international law. It is the responsibility of the Security Council to maintain international peace and security. So I don't think that we need nuclear-powered submarines. Insofar as the Arctic is concerned—please do not misunderstand me, I am not making a general statement with respect to defence policy for the East and the West coasts, which is outside my field of competence—I rather doubt it.

But to come back specifically to your question, that is, what I mean by effective control. I do believe that a Class 8 icebreaker, which would permit us to exercise surveillance over those waters year-round, except for the McClure Strait where you would need a Class 10, is the minimum we need.

Gwynne Dyer
Can I follow up, Professor Pharand, on the question of the nuclear-free zone in the Arctic that you are speaking of? The one difficulty that invariably comes up when one discusses a nuclear-free zone in the Arctic Basin is that, while neither Canada nor the United States has indispensable or even major military installations fixed in the area—though the Americans obviously have nuclear submarines transiting the area—the Soviet Union has its major naval base in the Arctic and would be required to make a quite disproportionate sacrifice in the event of agreement on a nuclear-free zone. Do you see any way around that?

Pharand
No, I really do not, but I am very pleased to answer the question. I think that, if we are serious about having a nuclear-free zone in the Arctic, it

Adrienne Clarkson is executive producer of Adrienne Clarkson's Summer Festival, *CBC Television, and publisher of Adrienne Clarkson Books, McClelland & Stewart. She is also a writer, lecturer and public speaker. She holds bachelor's and master's degrees in English literature from Trinity College, University of Toronto, and has studied French at the Sorbonne. She returned to Canada in March 1987 after two years in Paris as Ontario's Agent General. Prior to that she was a national personality, starring in such CBC public affairs programs as* Take Thirty, Adrienne at Large *and* the fifth estate. *She has received four ACTRA awards and her programs have been recognized by two Emmy awards. She is the recipient of two honorary degrees and is a Senior Fellow of Massey College at the University of Toronto.*

Question Session

has to include the Kola Peninsula where the Soviet Union, with its Third Fleet, has the highest concentration of nuclear submarines, probably in the world—I am not absolutely sure of that, but it certainly has a very high concentration indeed. Yes, I do believe that it would have to include that. However, in the speech by General Secretary, now President, Gorbachev on October 1, 1987, in Murmansk, he did say, if my recollection is correct, that the Soviet Union would go so far as to consider at least diminishing the Kola Peninsula fleet. I don't think that the difficulty is so great as to be beyond at least trying and seriously considering the matter.

Ann Medina
There is a feeling I get when we start talking in the world of international diplomacy that I am very puzzled.

Pharand
Excuse me, please do not misunderstand me. I am not an international diplomat; far from it. I would never have qualified.

Stephen Lewis
I don't know that that is true. Almost anyone can qualify these days.

Medina
Actually I am directing the question to another noninternational diplomat beside you, Professor—Mr. Slipchenko.

We have heard for so long about the urgency of dealing with environmental matters; we have been hearing for so long about the necessity of facing head-on some of the problems of nuclear proliferation, and yet when we get into international diplomacy, I see it as a kind of fantasy land. We heard from the Finnish ambassador that finally there will be some kind of an international cooperative study of environmental impact. We have heard that finally the Canadians and the Soviets have a program for the exchange of scientific ideas. And we are supposed to be ecstatic that such progress has been made, despite hearing about how urgent problems have been. I detected at the beginning of your speech, Mr. Slipchenko, a feeling of perhaps frustration. Perhaps you could tell us of some of the obstacles you have discovered over these thirty years, and why the world of international diplomacy has not by now gone a little further. What kinds of obstacles have you seen? What kinds of frustrations have you felt?

Walter Slipchenko
I am not a diplomat, being someone who was involved in the day-to-day

operation of trying to institute a program of exchanges. In terms of frustration, we started—and that includes our Soviet colleagues too—discussing this interchange or agreement in 1971. We signed one in 1984. So the first thing one learns is that one has to have patience in dealing with this. The second point you raised—or so I gather—is: Is this really important?

Medina
It is obviously making an important step, but is it as much as we really could have achieved? Or is it enough? You have the answers.

Slipchenko
I would say that it is never enough. This is the basic problem. On the other hand, the trust that is being built up between ourselves and our Soviet colleagues is a two-way trust. I remember in the first negotiation we said that, if it was going to be an Arctic science exchange, we had to go into the Soviet North. And the answer was, Why? Why not just come to Moscow and we will discuss the program of Arctic science exchanges? Now remember that this was the year 1974. It was what I call BG—Before Gorbachev. It was a year when the Soviets neither had social problems nor did they have pollution problems; these were primarily capitalist phenomena. Since then, of course, we have discussed these things. It has been very important, as I say, for this trust to be built up over a number of years, and the trust was based on the fact that the Canadians were not interested in looking at developments in the Soviet Union from a negative aspect. We have enough that is negative in Canada, in northern Canada. What we were looking for was to find out what they had done that we could learn from them. This is why this trust has taken so many years. I think, in light of a number of international events, that it probably could not have been achieved by the end of 1979; and then, as you know, there was Afghanistan in between that threw everything out of kilter. The process has been a long process. It has been built on an agreement on both sides, and we both agree that there are things to learn from each other.

Medina
Round two I guess. Going back to something that was talked about this morning by the Minister, Mary Collins, and also going back to your comment at the beginning that you were awaiting a policy from the Canadian government, an Arctic policy, a distinct, coherent, strong policy—will there be one? Will it be something that says it will be tied to building, strengthening East-West relations? Do you realistically expect such a policy?

Slipchenko

I do expect one. What is happening now is that, every time a Canadian goes out in any area, he or she is making policy. And you now have the regional governments—territorial governments—having international relationships; for example, the relationship of the government of the Northwest Territories with Greenland, Alaska, Norway and, of course, the Soviet Union; and so policy is being established. I think this is going to force the federal government into setting some general guidelines. It will be very important that they do it quickly, so that other people will know in just what framework we should be operating.

Dyer

Mr. Slipchenko, we were talking this morning at considerable length about the Inuit Circumpolar Conference and the developing links between the Inuit and, to a lesser extent, other native people in the polar basin. To what extent does the territorial status of the Northwest Territories help or hinder the institutionalization of those things? I am having difficulty asking this question exactly, but my perception is that in some ways these now-informal links between Greenland Inuit, Canadian Inuit, Alaskan Inuit and, as of this year, Siberian Inuit may eventually and to everybody's benefit take on some more regular institutional form within the various existing territorial-state entities. Has anybody in the Northwest Territories government given any thought to this? Has there been any discussion about it? Does the territorial status, as opposed to provincial status, aid or hinder this?

Slipchenko

Generally, I can answer that both federally and territorially there has always been support for the Inuit Circumpolar Conference. In fact, on all occasions before recognition was given to the participation of Soviet Yuit at this next conference in Sisimiut,[1] there were a number of presentations on behalf of the ICC by, as I said, both territorial officials and federal officials.

As to a territorial policy of the kind you mention, I don't believe there is one at the present time. I don't think there is need for one because there is support for a pan-Inuit organization, remembering now that the ICC does not represent all the polar people. You have throughout the Soviet North another eight separate indigenous peoples who are not related to the Inuit, one being the Sami. You also have nine time zones throughout which they live. So I can say now that there is no policy. The only policy that I have seen is supporting the ICC.

Hughes
Mr. Slipchenko, you talked in quite a bit of detail about the bilateral agreement with the Soviet Union. I wonder if you could just speak briefly about the agreement we have with the United States. Is it a similar kind of agreement, and how does the Soviet agreement impact on the separate United States–Canada agreement?

Slipchenko
First of all, what we have with the Soviet Union is a working program. We don't have an agreement with the Soviet Union on the Arctic. The program we have is a specific program covering a number of subject areas. This is why I argued for the necessity of a broader agreement with the Soviet Union.

Concerning the United States, again it is out of my expertise. The agreement, it can be said, is a general agreement helping certain areas, such as research and the passage of icebreakers. But generally speaking, you can't compare the two. It is like comparing apples and oranges. They are entirely different.

Pharand
I would like to add just a word on this "Arctic cooperation agreement," as it is called, between Canada and the United States. First of all, the intention was to have an agreement whereby the United States was to recognize Canada's claim to those waters as being internal in return for Canada giving the United States right of passage. However, that agreement did not come to pass. First, this agreement, as just said by Walter, applies to icebreakers only. Second, they will ask permission, yes, and they did ask permission on two occasions last year for their icebreakers to enter Canadian waters. However, there is a notwithstanding clause, I would call it, at the very end of the agreement, that says that, regardless of the provisions of the agreement or, indeed, the practice implementing the agreement, the respective positions of the parties will remain the same—which is much worse. That is, we will continue to say they are internal waters, and they will continue to say they are international waters.

Lewis
Sounds a little like the American position on the Free Trade Agreement.

Adrienne Clarkson[2]
I think it is wonderful to come to a place like Edmonton where there are so many people who are interested in this. My only regret is that this is not happening down in Sodom and Gomorrah in central Canada, because

Dyer
This question is for General Greenaway. In his discussion of how we dealt with air control and air surveillance and how we might do it better, my understanding was that we could cover it all by ourselves if we only integrated the military and the civil systems we have. I have heard many times that, if we were not as closely linked with the Americans as we are now in a military sense, we would not be able to ensure our own security, ensure the surveillance of our own territory. We are critically and inevitably dependent on them somehow to do it for us. I understood you to say that if we actually spent our money more wisely and forced the Department of Transport and the Department of National Defence to cooperate and integrate their systems, we could indeed provide our own territorial surveillance, airspace and air traffic control out of our own pocket without beggaring ourselves. Am I correct in that understanding?

Keith R. Greenaway
Yes, although the standard answer is that it would be too costly. But this answer is not correct. It is absolutely not correct; we can do it. But you must keep in mind that there are two major fiefdoms here at odds, with control, and it is very difficult to bring them together. It is being done in other countries: Australia is moving in this direction; France has moved. The Americans did it very quietly by putting their radars around their country in a joint civil-military system.

You can take the military side of it and carry it further. If you want certainty in following an intruder whose intrusion is offensive in nature—which could be a strike—you need continuous surveillance with an intercept capability. This is of course where AWACS[3] can come in, and this is where joint operations come in, in a period of tension. But for the day-to-day surveillance of aircraft going through our airspace, yes, it is more economical to do it the way I have said. It can be done; I just don't buy it, and others that look into this do not buy it, that it is too costly.

Hughes
I want to ask a question of Mr. Robertson. You said that the federal government—and also the territorial government, but it is the federal government that I am interested in—has lacked the political will to push for the creation of Nunavut. I am wondering why you think that is, and what is the downside for the federal government of splitting the Northwest Territories?

Gordon Robertson
The evidence for the lack of will is that they have taken no position to help bring about an agreement, as far as I am aware and I think I am correct in that. What is the downside? I myself don't see any downside at all. I suspect that the main reason for not having done more than they have is that it does involve taking a position where it is more comfortable not to take a position and the political benefit for taking it may not be obvious. And if the political benefit is not obvious and there is no significant downside, why make it uncomfortable? I suspect it is not a great deal more complex than that. I can't answer your question beyond that. I don't know of any reason why they should not have taken a definite position.

I should go back into what's really quite ancient history and say that the council of the Northwest Territories as far back as 1960 recommended in favour of division of the Northwest Territories. It got very close but it didn't proceed at that time. Then it was revived, as I mentioned in my talk, in the 1970s by the Inuit themselves. I suspect that an additional factor in the Northwest Territories may be that the non-native group finds it more comfortable and convenient to leave the situation as it is.

I think also that the thesis has been developed, and to some degree sold, that devolution of powers is the same as aboriginal self-government, or the equivalent of it. It is not. At the present time, in the political enactments that are relevant for the Northwest Territories, there is no protection whatever for the native groups and native peoples, which would be one of the essentials, I think, if the Northwest Territories were divided.

Gordon Wade (Edson District Oldtimers Association, Edson, Alberta)
I would like to direct my question to Mr. Greenaway. Welcome to Alberta, Mr. Greenaway. Mr. Greenaway used to be my commanding officer twenty years ago. You spoke of air surveillance in the Arctic. What are your thoughts on Canada's need for nuclear submarines in the Arctic? Aren't there more practical and economical methods for Canada's surveillance of these waters?

Greenaway
I'm not a sailor but I won't dodge the question. I think that if you decide that you must have subsurface surveillance—and keep in mind that the most logical movement of intruders that do not want to be observed will be subsurface—the only way you can monitor them along the two thousand kilometre coastline, along the perimeter, not within the Archipelago but outside it, with present technology is by a nuclear-powered submarine.

This is not taking into account requirements in the Pacific or the Atlantic. As Professor Pharand said, and there is no question about it, nuclear power seems to be the kind that most maritime thinkers favour. If you want to monitor that subsurface area, you require the nuclear-powered submarine. In the channels, you can use detection systems at choke points. Because detection systems are static and because naval and monitoring tactical considerations may mean that you should follow the intruders, nuclear power is required for doing so.

Wendy Wright (Toronto Disarmament Network, Ontario)
My question is for Gordon Robertson. I am wondering what your opinion of the Meech Lake Accord is, because of the lack of representation by the territories and in relation to what you were saying about the formation of a third territory in the North.

Robertson
As far as the process of Meech Lake is concerned—that is, the process by which the agreement was worked out—I can understand very well the irritation and embarrassment and anger of the people in the territories at the fact that they had no representation whatever in the process. Apart from the process, if one looks at the substance of Meech Lake, what has caused the greatest wrath in the North is the fact that, if Meech Lake comes into effect, in future there will have to be unanimous consent of parliament and the legislatures of the ten provinces to establish new provinces in the territories. In terms of substance, I think that is the main source of anger in the territories.

I have some difficulty in sharing fully the wrath over that because I wrote a short book, the thesis of which is not very popular in the Northwest Territories, but is, I am afraid, correct. The basic thesis is that the means of financing provinces is such that it will not be possible to have provinces in the northern territories. The basic difficulty—I can't go into the complexities of it—is that the North is a very, very high cost area, and because of the way in which equalization payments are calculated, the North wouldn't qualify for equalization and for financial support. But those are the main reasons why there is difficulty about Meech Lake.

Walter Doskoch
Prior to my question I would like to make a few remarks. I have lived in this country for over sixty years and have found in those over sixty years that the fight to be objective is the most important fight mankind has. So in the conclusion that I have drawn listening to this debate this afternoon I have found this out: that if politicians and generals got paid what they were worth they would die of starvation on welfare. That is the first comment. My second comment is: the North is loaded with Inuit people,

people who have been there for thousands and thousands and thousands of years. Those people live in many countries. Why doesn't this group who say "The True North Strong and Free" say, "Listen, you politicians, the North belongs to the Inuit, the raw materials, the gas, the gold, all the precious metals, the oil belongs to those people"? Why doesn't this meeting say, "Let them have that oil and let the Inuit develop it for themselves and not us"? Why do we allow Alberta to develop paper mills that will contaminate the Athabasca River from north of Edmonton right to the Arctic, contaminating people, animals and you name it?

Lance McFadzen (M. E. Lazerte High School, Edmonton, Alberta)
I would like to address my question to General Greenaway. Sir, you have stated that our surveillance of our Arctic airspace is inadequate, and I agree. However, even if we could detect hostile intruders, we could do nothing about them. In Canada's attempt to use the CF-18 Hornet as the jack-of-all-trades of air defence, we are shooting ourselves in the foot. While an excellent ground attack fighter-bomber, the Hornet's comparatively sluggish top speed of Mach 1.8 lacks the speed of an interceptor, and its endurance is not that of an air superiority fighter. What do you recommend to resolve this problem?

Greenaway
I think the first thing is not to worry about shooting things down. It is a matter of detecting and identifying those that we don't know that are there, which may not be enemy aircraft at all. They may be just people snooping. We have a house here, but there are no doors on it, and we have no way of detecting who is going through. This is the problem. I don't think this is the place to go into the problem of shooting down, but the tactics are worked out for it. There is a certain degree of tension built up before that happens, and you do not normally, in our environment, shoot and ask questions after. You bring intrusions into the World Court, as Professor Pharand said. You highlight them, you point them out. It is a point in our North today, in this area where we have roughly six thousand aircraft a year going through. The reason I can say six thousand is because these are basically cooperative aircraft that file flight plans when they are flying over our northern area. But we don't know how many come through the area that do not file flight plans. I think the fundamental problem is getting the surveillance in place and the technology in place to monitor it. I think the fighting is another story entirely.

Clarkson
And the specific question about the CF-18; you are happy with it?

Question Session

Greenaway
Yes. The CF-18 is a very good aircraft. I think the military did an excellent job in deciding on it. There is a lot of growth in that aircraft, and it is a good aircraft, a flexible aircraft, so that if our roles in the various alliances change, that aircraft can be adjusted to the different roles.

Garnet Thomas
My question is for General Greenaway, and I hope General Huddleston will perhaps address it tomorrow as well. My question also relates to the CF-18 indirectly and to the cruise missile. We have heard recently with the latest tests that the Canadian pilots in the CF-18s, along with their American counterparts, are practising the interception of missiles. From my knowledge as a pilot—and I think you've confirmed this in talking about the weaknesses in our radar coverage—there is no effective way I know of to intercept not one cruise missile but hundreds, say, in an incoming attack, other than perhaps a technique that the Americans have developed, which is to lob nuclear weapons from fighters into the general pathway of incoming missiles. When did it become Canadian policy to begin using nuclear weapons on our fighters? Are they now being stored in northern bases and at Cold Lake? If we are not using this technique, if there is some other effective way that I don't know about, could you perhaps elaborate in some general way?

Greenaway
Moderator, I believe that question should be left to General Huddleston tomorrow, who would be in a much better position than I am to answer that question. The one thing I would say is that these cruise missile tests provide our military with excellent opportunities to view missiles in flight and to gain knowledge of how the missiles operate. I'm not saying that that is the best place for testing, but just that our military does get a chance to look at these things, which, regardless of what we say, are going to be in the inventories. We are living in the real world, and they are there. It would be very foolish if our military didn't know something about them.

Quentin Miciak
Professor Pharand and General Greenaway, both of you gentlemen, when you were talking about detecting intruders, be it in our waters or airways, referred to these detections of intrusion being turned over to international bodies like the Security Council or the World Court. How effective have these international bodies been in dealing with intrusions into national waterways or airspaces in the past?

Pharand
First, it is not very often possible to take such a case of intrusion or a similar one to the International Court of Justice for the simple reason that the jurisdiction of the Court is not compulsory and, since the United States withdrew its acceptance of the Court's jurisdiction two and a half years ago, there is only one major power left, the United Kingdom, that still has an acceptance of jurisdiction before the Court. This does not of course prevent any two parties from agreeing to go before the Court. For instance, the United States agreed about a year ago to go to court on a case with Italy, but it wasn't a similar case at all.

Second, the only international body that you can go to, I believe, is the Security Council. It is the primary function and responsibility of the Security Council of the United Nations to deal with any threat of the use of force. However, in answer to your question specifically, I have to admit that whenever there is a major power involved, one of the big five, it is not really possible for the Security Council to take any action. Why? Because they have what we call a veto power under Article 27 of the charter. So there is a certain flaw, insofar as the implementation of the collective security system provided in the United Nations charter is concerned. To answer your question even more specifically, yes, your implication is correct. The theory is sound, but the implementation, when the will of the states is not there, leaves very much to be desired and is sometimes impossible.

Gordon Scott (Calgary, Alberta)
I am addressing this question to Gordon Robertson and Donat Pharand. We have heard much about the difficulties that people have had in international negotiations, from the late nineteenth century right up until now. What I would like to ask you is whether, taking an overall view of that period of time, you see an acceleration of international cooperation and an increase in worldwide awareness and understanding of the necessity for all people to work in cooperation?

Pharand
I would answer your question in the affirmative. I think, particularly in the last couple of years, we have seen a much better international atmosphere for cooperation, and of course I am speaking particularly of cooperation on the part of the Soviet Union. After all, that is not, shall we say, traditional. It has not been traditional for that superpower to come forth as it has now done, saying it is willing to cooperate both bilaterally and multilaterally. This is the new part. It seems to me that it has happened only in the last couple of years, beginning with President Gorbachev's regime.

I think we have to take seriously that expression of willingness to cooperate, and we have to maximize it to the best of our ability, to really test it to the limit, as it were. If we cooperate in non-military matters, matters of common interest, such as the protection of the marine environment, the environment generally and scientific research, there is a good chance that we are going to continue in that route of cooperation and perhaps eventually come to some meaningful agreement on arms, not only limitation and control but eventual disarmament. So I say yes, I think that there is some hope there.

Robertson

I would agree completely with everything that Professor Pharand has said. There are just one or two things I might add. It is easy to become discouraged at times at the weakness of the arrangements that we have for multilateral international cooperation, for international cooperation in general. But I think we should remember that we didn't have any arrangements at all for multilateral, international cooperation of a permanent or standing kind until after the First World War. That was when, for the first time, a League of Nations was established in order to establish some sort of regime of international cooperation and law. It was flawed from the beginning because the United States was not a member of it, and it failed, of course, in the 1930s, before the Second World War.

The United Nations, whatever its problems may be, has worked far, far better than the League of Nations did and has worked for far longer; and we do have cooperative regimes and regimes of international harmony and law in a much wider area of activity than we ever had before in the world. So I think we should not become too discouraged. In addition to that, as Professor Pharand has said, we have recently, in the last two years, seen a major change that perhaps is going to usher in an even better period. So I would say definitely it is better than it was.

Jim Lavers

The question is to Mr. Robertson or any of the other panelists. I am going to give you a general question first: Which would you consider more important, political viability or economic viability in the North? I ask that because, based on what you said about the present scheme, it would be impossible for the North ever to become viable because of the existing grant structure.

I am also asking a specific question, which I would like you or anyone else to answer. To what extent do you consider the energy aspects, that is, pipelines, and therefore American interests and influence, to be part of the reason for the apathy of the present government in not moving towards provincehood for the various parts of the North?

Robertson

On the first question, about relative importance and the possibility of political versus economic development for the territories, I would refer to what Tom Berger said this morning about the unlikelihood of our ever being able to provide in the Northwest Territories—and perhaps it is the same in Yukon, I'm not quite so sure there—a job for everyone of the kind that we think of as the result, the hoped-for result, of economic development. I don't think that we will see that in the Northwest Territories, because the problems of the Arctic especially, but also of the Subarctic, are simply too intractable, too great, too difficult. And I agree completely with what Mr. Berger said today about the acceptability—the inevitability I think, but certainly the acceptability—of considering a situation in which we simply recognize that the North is going to require, as many areas of the country have required, a fair amount of subsidization in order to be economically viable in the long term. And I don't think there is anything wrong with that. We basically have to decide, I think, whether we want to have a population in the North that is as self-respecting, self-governing and as self-supporting as it can be or whether we don't. If we do want it, I think we have to accept a fair degree of subsidization for quite a long time, and I see no reason why we should not do that. The population is not large, it would not be vastly expensive for a country of 26 million people.

Mini Freeman

If I may I would like to make a statement and ask a question. I was bothered by Gordon Robertson's talk earlier. He painted Inuit people as being alcoholics and full of drug abuse, using the excuse that their learning of southern ways is all mixed up. That excuse has been used so long; it's been fifteen years since it started. Today Inuit people have learned a great deal about their mistakes and are learning on their own how to use southern material, if you like. So why don't people, instead of thinking that we are all alcoholics and that we don't care what is happening on our land, why don't people go up there and see?

Clarkson

I think, Mini, in all fairness to Mr. Robertson's statement, he didn't point a finger, he simply was making a kind of analysis of certain problems that have arisen since southern values and a southern work force have gone north.

Freeman

I guess it is because I am from the North and it so deeply touched me that it sounded like that. My question is—it is so obvious—why are there no other native people here? It is their land and they are affected. There are the Inuit Tapirisat of Canada, the Inuit Broadcasting Corporation and so on. You name it, they are all organizations of Inuit people and they are not here. I only see one. Does anyone have an answer to that?

Clarkson
It is a very important question and I think it is one that perhaps we can all address together; maybe we can talk about that again tomorrow. I don't think this particular panel can answer that for you, but I think the point is well taken, Mini. Thank you.

Questioner
I came to this inquiry to learn about two issues, peace and security. I'm a little bit disappointed. We have heard so much about security and so little about coexistence and peace. My feeling is, if we had more peace, we would need less security, and that's what we should be striving for. I would hope and ask that tomorrow, those who participate will emphasize the peace part of this inquiry, so that we can proceed to lessen the need for security.

Clarkson
Thank you for that statement. I must say that I think the analyses, particularly of Professor Pharand and Mr. Robertson, have addressed the problem of peace through rather complex and very worthwhile means. I think, although all of us want to believe that we can have a peaceful world, very simply it doesn't work that way. I am very appreciative of the analyses we have heard—of a situation very complicated internationally and legalistically—that can help us understand what it takes to make a peaceful situation. So although I appreciate the point, I don't think the criticism should be as strong as that, if I may make that interjection.

Detlef Taritz
First of all, do you think the Soviet Union is a threat to Arctic sovereignty? Do you feel the United States is a threat? And what are you going to do, blow them both out of the water, or merely inform Brian Mulroney that they are there?

Greenaway
From a sovereignty point of view, I think we don't worry about the nationality. It is anyone; we need to know who comes in the door and be able to identify them. It could be an American aircraft, it could be a Soviet aircraft, it could be a French aircraft, it could be a British aircraft, it could be any one of them. But we haven't a mechanism by which they

can unlatch the door and come in and identify themselves; that's really what it is. I'm not saying that there is a threat to us from any aircraft. It is not one country any more than the other.

Pharand
I would just make one little comment here editorially, that I think it very fascinating, that whenever we talk about sovereignty in this country, we always say we are nationalists. There is no other country in the world where, when people talk about their country and what it means, they call themselves nationalists. I've never heard French persons call themselves French nationalists, or Americans call themselves American nationalists. It's just an interesting point actually. I think in Canada you could be a Canadian, couldn't you?

Johann Luxen (Edmonton, Alberta)
I have a question for General Greenaway. How does one cover a radiation leak in a submarine at the bottom of the sea with concrete, as they did in Chernobyl, to stop that radiation? Whether the submarine got there by accident or enemy action is immaterial to me.

Greenaway
I think you are referring to the risk of nuclear power. I think the navies of the world that have used nuclear-powered ships and submarines have a very enviable record. They have been using them for forty years or so without any serious accident of any type, not as serious as the oil pollution off the West Coast here just the other day or the other one down on the East Coast. The navies and the coastguards too have had very good experience with nuclear-powered ships and submarines. That is all I can say about it—I'm no expert in it, but I know the record is good.

Questioner
Excuse me, sir. A spaceship blew up too and it was supposed to be safe. And do you think just because there might not be an accident, they shouldn't have some way or means of dealing with that thing?

Greenaway
I don't think you can compare the Chernobyl incident with the technology in the nuclear-powered submarines of either the Soviet fleet or the American fleet or the British or the French fleets. I don't think you can do a comparison. The technologies are not of the same vintage by any means. They are of different vintage with different controls, which are much more rigid in the military than at Chernobyl.

Question Session

Mary Ann Ashford (Doctors for Peace)
I couldn't resist standing on a point of information. There are four nuclear-powered submarines on the floor of the ocean. They went down with all hands. Surely those are serious accidents.

Greenaway
I might add that there was nothing wrong with the nuclear power plants of those four submarines. They foundered for other reasons. There has been no leakage at all; they are being monitored.

Notes

1. Sisimiut is the Greenlandic name for Holsteinsborg, a town on the west coast of Greenland. Greenlandic place names are gradually replacing European ones. (Editor)
2. Adrienne Clarkson took over the role of moderator from Stephen Lewis for the rest of this question session.
3. The Airborne Warning and Control System of the U.S. Air Force.

CHAPTER 14
The Arctic, Northern Waters and Arms Control

JOHAN JØRGEN HOLST

I shall speak to you briefly this morning on the subject of the Arctic, northern waters and arms control. And I must warn you that because the program is fairly dense, so is the argument.

The Arctic links the security of Europe with the security of North America. Its strategic significance derives from the East-West competition in relation to the central balance of nuclear deterrence as well as to the continental balance of power in Europe. Security in the Arctic is linked inextricably to the structure and process of security arrangements between the two alliances that emerged in the wake of the controversy of the outcome of the Second World War.

A stabilization of the central balance of nuclear deterrence would tend to reduce the pressures of competition affecting the policies of the major powers in regard to developments in the Arctic. Several factors determine the strategic perspectives that now point in the direction of enhanced stability.

First of all, the principal powers have discovered the limited convertibility of nuclear forces into politically useful currency. Their function is confined very largely to mutual denial. Attempting to compel others through nuclear "diplomacy" has not provided credible options. Neither side can realistically expect to acquire meaningful superiority within the system of nuclear deterrence. Hence, the principal powers now share an interest in limiting rather than expanding the nuclear competition. They recognize that the very competition harbours the danger of inadvertent conflict, which could result in catastrophe for both sides. Hence, the principal powers share an interest also in stabilizing their competition so as to prevent concerns about the balance from pushing them to points of no re-

Johan Jørgen Holst

turn. It is the dangers of August 1914, rather than of September 1939, that loom on the horizon.

Secondly, the unique quality of nuclear weapons has penetrated the moral consciousness of humankind. Nuclear weapons are different from other means of destruction, not only because of their capacity for instant and extensive destruction but because of their largely unpredictable genetic and ecological consequences; they may destroy the conditions of life as we know them for future generations. Hence, the consequences of nuclear war would not be confined to the distribution of power and influence in relations among states and nations but would extend rather to the very essence of human life. Since nuclear weapons do not lend themselves to "disinvention," and since nuclear deterrence cannot be made foolproof, states and nations cannot escape from the imperative of minimizing the danger of nuclear war.

Thirdly, stability, in the sense of low expected premiums from first strikes, requires careful effort and cannot be taken for granted. Technological developments create changing requirements for deterrence. The system of nuclear deterrence does not define the ultimate means for the preservation of peace. Rather, it contributes a temporary expedient in regard to both moral imperatives and practical opportunities. However, it cannot be transcended by unilateral means or technological manipulation. Such attempts are likely to generate more competition and less stability. Orderly transition to security arrangements beyond deterrence can only be

Johan Jørgen Holst is Norway's Minister of Defence. He has served in previous Norwegian governments as state secretary for defence and state secretary for foreign affairs. He is a graduate of the Norwegian Army Language School, in Russian, and of Columbia University, in government; he also has a master's degree in political science from the University of Oslo. He was the scientific consultant to the Palme Commission on Disarmament and International Security Issues and special advisor to the chairman of the World Commission on Environment and Development. He is a member of the international commission of the Norwegian Labour Party as well as a member of the council of the International Institute for Strategic Studies. His career has led him from research roles with the Center for International Affairs, Harvard University; the Norwegian Defence Research Establishment; and the Hudson Institute, to the position of director of the Norwegian Institute of International Affairs and membership in such bodies as the Norwegian government's advisory council on Arms Control and Disarmament and the Trilateral Commission's task force on ocean policy. Prior to his appointment as minister, he was a member of many influential boards and a frequent speaker and author, mainly on security and foreign policy.

envisaged as a cooperative undertaking reflecting a shared conceptual framework of common security. Progress is likely to be incremental rather than systemic.

Fourthly, the Arctic will continue to provide the most direct avenue of approach for strategic weapons travelling between the heartlands of the two principal powers. By extension the Arctic will constitute a forward area of warning and defence against attacks by such weapons. Furthermore, it will provide patrol areas for submarine-based strategic systems. Protection, surveillance and challenge of such patrols will define some of the major tasks for the navies of the principal powers. Rules of engagement and disengagement in that context will impact on the political position of the littoral states.

The Arctic Ocean is a large sea with relatively narrow outlets to the oceans to the south. It has not played a major role in past conflicts among nations. However, its strategic importance has grown with the advent of nuclear-powered, missile-carrying strategic submarines. Increasingly, such weapon systems have been able to penetrate and exploit inner space below the Arctic ice. Patrols by strategic-missile submarines constitute both potential targets and actual objects of protection for the naval forces of the major powers, particularly ocean-going attack submarines. The search for stability inevitably raises the issue of naval arms control. Several observations need to be made about the essence of naval forces before the potential for naval arms control can be assessed properly.

Firstly, naval forces constitute essentially mobile military capabilities. Their reach is global. Hence, they do not lend themselves very easily to regional limitation. Regional naval limitation regimes are likely to prove unstable, as they would inherently be vulnerable to destruction by naval forces from outside the region.

Secondly, naval forces constitute essentially multiple-mission military capabilities. They can be used in a variety of roles: to fight other ships for command of the seas; to chase and destroy submarines or surface vessels, including merchant vessels; to bombard targets on land; or to provide protection to defending forces on land.

Thirdly, naval forces constitute potential instruments for political influence. Enjoying freedom of navigation on the high seas, they cast political shadows before them, particularly onto the shores of the littoral states. However, since the dependence of nations on supplies by sea varies considerably, symmetric limitations on access to particular ocean areas could have asymmetric political effects.

Fourthly, naval forces constitute military instruments that are deployed and operate largely outside the areas of jurisdiction of the nation-states. Therefore, the regulatory powers of the coastal states constrain the freedom of manoeuvre of the flag state but in rather marginal ways. Through centuries, sailors have had to develop rules of the road so as to reduce the danger of incidents at sea and their possible escalation to armed conflict.

Several conclusions suggest themselves concerning approaches to naval arms control that are consistent with the essential character of naval forces. Firstly, as a general rule, limitation should be *global* rather than regional in scope.

Secondly, limitations should focus on the *inventories* of specific types of naval forces rather than on missions, since the latter are typically conducted by a variety of forces in a variety of combinations. In many instances, complete elimination rather than limitation by agreed ceilings would provide more stable regimes, particularly from the point of view of verification.

Thirdly, naval arms control must be considered in a *comprehensive strategic context,* taking into account the relative dependence of nations on interior overland lines of supply and communication, on the one hand, and exterior overseas lines of supply and communication on the other.

Fourthly, *confidence-building measures* at sea should take into account the specific nature of naval operations, the navigational traditions that have developed over the years, the principle of the freedom of the seas and the perspective of mutual advantage.

When considering the current agenda of arms control and how it could affect developments in the Arctic, several propositions may be advanced. Firstly, a strategic arms reduction agreement based on a fifty percent cut would likely result in a reduction in the number of strategic nuclear submarines operating in the Arctic and near-Arctic oceans. However, specific constraints on destabilizing, heavy, land-based missiles and an emphasis on survivability could cause a relative shift from land-based systems to sea-based systems and from large strategic submarines with many missiles to smaller submarines with fewer missiles.

Stability could be enhanced also by reducing the ratio of warheads to launchers, thus moving away from systems with multiple independently targetable re-entry vehicles (MIRVs) and towards single-warhead systems. A Strategic Arms Reduction Talks (START) agreement could affect both the size and the structure of the strategic nuclear forces maintained by the two principal powers. It would in all likelihood cause the Soviets to retire their strategic submarines with intermediate-range nuclear missiles presently patrolling the Norwegian Sea, and presumably covering targets in Western Europe, in the wake of the Intermediate Nuclear Forces Treaty[1] prohibiting the deployment of intermediate-range missiles on land.

The Soviet choice of a northern patrol option in or near the Arctic for their long-range strategic submarine-launched ballistic missile (SLBM) force would essentially create a situation in which they would threaten targets in Western Europe and North America from the same positions and with the same systems, thus contributing, paradoxically, to the strategic unity of the North Atlantic Treaty Organization (NATO). However, a Soviet choice of short-range, short-time-of-flight, depressed SLBM trajectories would require forward patrols off the coast of North America. This choice would constitute a potential first-strike threat against the land-based components of the U.S. strategic deterrent. In order to enhance stability, the testing and deployment of such SLBMs should be prohibited in the START agreement. The long-range Soviet northern patrol option is, in my view, more consistent with stability.

Long-range, nuclear-tipped sea-launched cruise missiles (SLCMs) are operational with the navies of both of the principal powers. In addition, the Soviet Union is developing a supersonic version. From a Norwegian perspective, this development is a matter of considerable concern. It threatens to redirect the nuclear arms competition to northern waters, and it may constitute a challenge to the condition of low tension that has prevailed in the European North. The impact might be particularly cumbersome in the context of progress in détente and arms control on the ground in Europe.

There are many more potential targets for attacks with nuclear-tipped submarine-launched cruise missiles near the coasts of North America and Western Europe than there are near the coasts of the Soviet Union, and they are vulnerable because of the short flight times from Soviet surface vessels and submarines. Norway, you should remember, has a very extended coastline and an extremely shallow territory relative to its coast. The geographical asymmetry in this respect suggests a strong Western interest in prohibiting nuclear SLCMs.

Furthermore, SLCMs could introduce pre-emptive instabilities in a conventional crisis to which naval forces would be committed, as nuclear cruise missiles would be proliferated on a variety of ships. This, in turn, could reduce the flexibility of the American navy and extend conventional deterrence to the exposed areas of the European North, since vessels with SLCMs might be withheld for political reasons.

A ban on nuclear-tipped SLCMs would probably depend to a large extent on whether the principal powers are able to conclude a START agreement and develop force postures with survivable land-based and sea-based ballistic missiles.

In the absence of such a regime, SLCMs may seem necessary in order to dissuade attacks by multiplying the targets that an opponent would have to eliminate in a first strike. Furthermore, Norway's interests are here potentially at variance with those of some of our continental NATO partners, who may view SLCMs as a substitute for land-based intermediate-range nuclear forces. However, in the event that East and West should agree on substantial reductions of levels of conventional forces assembled on the continent of Europe—and on their reconfiguration through preferential reductions of capacities for surprise attack, sustained offensive and seizing territory—the presumed requirements of such nuclear-strike options would diminish. Furthermore, a ban on nuclear SLCMs would presumably be easier to verify than a higher ceiling.

A more radical solution would involve the elimination of all nuclear weapons at sea except on missiles in strategic submarines. Such a radical solution could enhance strategic stability also by eliminating a naval nuclear threat to the survivability of strategic missile submarines.

The idea of a sanctuary for strategic missile submarines has sometimes been suggested as an arms control measure. However, the monitoring of such sanctuaries would be extremely difficult. It would require extensive cooperation between the two principal powers, and such cooperation could easily translate into claims for preferential rights in the ocean areas in question. It could also cast political shadows onto the shores of the littoral states, whose security and sovereignty would become closely entangled with the management and central balance of nuclear deterrence between the two principal powers. Limits on the deployment of fixed sensors would have but a marginal impact in the context of the quiet submarines, which we now see emerging.

Since the principal threat to nuclear-powered strategic ballistic-missile submarines (SSBNs) is mounted by nuclear-powered attack submarines (SSNs), it is possible to envisage stabilization through agreement on the ratio of each side's number of SSNs to the other's SSBNs. However, as such SSNs also constitute threats against the sea lines of communication linking, for example, North America and Western Europe, incentives would be present for NATO to use SSNs to exert pressure on the patrol areas of Soviet SSBNs. This would force the Soviets to deploy SSNs for the protection of the SSBNs, and thus prevent the SSNs from threatening the sea lines of communication. Hence, in the context of seeking increased stability both within the central balance of nuclear deterrence and in relation to the conventional balance in Europe, a ban on ocean-going SSNs could emerge as a promising option.

We may conclude, then, that the two principal candidates for possible elimination from the inventories of naval forces would seem to be the nuclear long-range sea-launched cruise missile and the ocean-going attack submarine. The principal candidates for limitation would be submarine-launched ballistic missiles and strategic missile submarines, as part of a comprehensive agreement for strategic arms reduction.

The Soviets have proposed measures that would limit the access of naval forces to northern waters off Europe. However, the impact of such arrangements, even if they were to be symmetrical, would have highly asymmetric implications.

The Soviet Northern Fleet is "home-ported" in the area. The American navy is "home-ported" in Virginia. Furthermore, Norway depends on allied reinforcements in an emergency. Such reinforcements would in large measure have to come by sea. Furthermore, NATO, as an alliance, depends on reinforcements from North America to Western Europe. Such reinforcements depend on the transatlantic sea lines of communication. The transatlantic sea lines of communication would be in serious jeopardy if Norwegian airfields were to be available to the Soviet Union or denied to the Western powers.

Norway's security and that of NATO as a whole depend on allied navies being able to protect the sea lines of communication in forward areas in the North and on extending protection to Norwegian territory against combined air-land-sea attacks, particularly in the high North. Hence, regional limitations on naval access would be likely to reduce stability rather than enhance it. In the event of an agreement on conventional stability in Europe involving reduced force levels in forward areas, NATO's dependence on the sea lines of communication would grow, as the role of capacities for reinforcements would grow in the overall East-West balance of forces.

A confidence-building regime at sea could be developed on the basis of the bilateral "incidents at sea" agreements, which have been concluded or are under negotiation between the Soviet Union on the one hand, and the United States, Great Britain, France and the Federal Republic of Germany on the other and which apply beyond the territorial seas. Similar agreements could be concluded between other pairs of interested parties. In addition, it is possible to envisage multilateral "incidents at sea" agreements entered into by the states that undertake naval or other activities involving the exploitation of ocean space in the northern waters and the Arctic. Special provisions may be needed to establish behavioural rules that can protect fragile and sensitive environments and regulate traffic in areas of dense activity associated with fishing, maritime traffic, naval operations and offshore activities.

The system of confidence-building measures that has been developed in the Conference on Security and Cooperation in Europe (CSCE), and amplified by the Conference on Disarmament in Europe (CDE) applies to military operations on land. These measures do not lend themselves to easy transfer to naval operations at sea. It is doubtful that such measures at sea should be confined to ocean areas in the vicinity of Europe rather than make up a more global regime.

However, certain elements of the CSCE/CDE confidence-building system that has been developed could be applied to the sea. The modalities would have to be carefully considered however. Provisions concerning notification of exercises would have to establish *the size* of notifiable exercises, measured by an agreed currency (manpower does not here seem particularly appropriate). Furthermore, the *delineation of exercise areas* would have to be defined in a more flexible manner than on land. *Observation* could take place from ships not directly participating in the exercise, combined perhaps with helicopter visits to participating units. The purpose would be to reduce the fear that routine exercises may contain the seeds of surprise attack, rather than to facilitate the gathering of intelligence.

Norway's strategic position is determined by its geographical location in general and by its proximity to Soviet forces and military installations on the Kola Peninsula in particular. The significance of geographical location is determined in large measure by technological developments and the constellation of international relations among the great powers. Vital strategic interests intersect in Norway's immediate vicinity. Military units that are operating or based in Norway's immediate proximity may be employed in the pursuit of interests in areas very distant from northern Europe. Conflicts in distant areas of the globe may thus affect Norwegian security. Norway can no longer enjoy security as a result of distance from the sources of international conflict or as a result of a peripheral position of marginal significance. Nor can Norway rely exclusively on the protective shields of friendly naval powers.

Traditionally, the northwestern region of Europe has been viewed as an isolated flank area relative to the central front. Limited war and *fait accompli* scenarios dominated security thinking during the 1960s and 1970s. However, during the 1980s, the North and the centre increasingly have come to be considered as an integral theatre from the point of view of military strategic planning. Developments in military technology, renewed attention to the problem of transatlantic reinforcements and changed maritime perspectives have led to a more holistic approach. We now view the defence of Norway in a European context and its strategic position in an Atlantic, and increasingly in an Arctic, perspective.

Soviet ground and air forces as well as naval infantry units on the Kola Peninsula have accumulated an increased capacity for rapid offensive action in the North. In this connection, the introduction of modern combat helicopters and the increased helicopter mobility of the ground troops are of particular significance and a source of worry to my country. The introduction of airborne command and control stations as well as modern fighter aircraft with a capacity comparable to that of Western aircraft has enabled the Soviet Union to enhance its capacity to conduct forward air operations.

The main forces on the Kola Peninsula are elements in the global competition between the Soviet Union on the one hand, and the United States and its Western allies on the other. They are not addressed to Norway specifically. However, they affect and complicate Norway's position.

In preparing for the newly initiated negotiations on Conventional Forces in Europe (CFE), Norway put particular emphasis on the need for a comprehensive rather than a regional approach. The negotiations should apply, in our view, to all of Europe from the Atlantic to the Urals, and from the Barents Sea to the Mediterranean. Norway argued strongly against subregional zones, as they would tend to weaken the links that sustain the security order in Europe and would expose the smaller, peripheral states to the local preponderance of the Soviet Union, which is, after all, the major military power on the Eurasian continent.

Special limitations may be required in the central area of maximum concentration of forces, and in subsequent phases, the system of variable geometry of limitations of particular forces that constitute special threats to the stability of certain sectors may be envisaged. But fixed zones for forced limitations should be avoided, in the Norwegian view. Norway was pleased, therefore, by the amplification that the unilateral-force cuts, which the Soviet Union has announced in order to reduce some of the salient asymmetries that weigh on the security order in Europe, should apply proportionately also to the Leningrad Military District, which includes the Kola Peninsula.

Norway has engaged in the careful construction of a *Nordpolitik* designed to promote peaceful, stabilizing and cooperative developments in the high North. It has both defence and civilian components.

The defence policy part of the Norwegian *Nordpolitik* proceeds from four main propositions. Firstly, Norway must cooperate with its allies, particularly with the United States, to ensure a capacity to control the ocean areas off northern Norway. The threshold may thereby be raised against direct attack, deterrence may be enhanced, the nuclear threshold may be raised and the position of the alliance may be improved.

Secondly, Norway must contribute to effective surveillance of naval activities in northern waters in order to obtain early warning and to prevent miscalculation. Reciprocally tolerated surveillance enables both sides to observe and demonstrate the absence of feared dispositions. Norway has assumed a major responsibility for surveillance of the most sensitive areas in the northeast.

Thirdly, occasional forward allied naval presence off northern Norway is important in order to provide credible protection for the transatlantic sea lines of communication. This does not imply preparations for horizontal escalation by Western naval forces in northern waters. The Soviet Union has too many comparative advantages in terms of general proximity and access as well as in terms of naval forces in the area. Furthermore, the dangers of explosive escalation are likely to seem forbidding to the West. However, a Western capacity for the projection of naval power into the northern waters is necessary in order to deter the Soviet Union from exploiting its local preponderance by resorting to horizontal escalation in response to a stalemated conflict elsewhere. Increased cooperation and arms control negotiations are likely to constrain such options further.

Fourthly, it is important to prevent an extension and intensification of the arms race as a result of competitive deployments of sea-launched cruise missiles in northern waters. Furthermore, Norway is interested in exploring the scope for appropriate confidence-building measures that could contribute to the prevention of incidents, inadvertent conflict and escalation at sea.

We should always recall, however, that genuine security involves a much broader agenda of cooperative undertakings to protect and preserve a fragile environment, promote a sustainable and equitable exploitation of natural resources, promote our understanding of the complex ecology of the far North, and so on. Progress in these areas would spin a network of cooperative threads across the divisions that have sustained and stimulated the military competition. Increased cooperation in the high North could reduce the saliency of the military conflict and its impact on the conduct of international relations by the states in the North. Arms control could gradually move the military effort away from the framework of competitive security to one of common security.

Notes

1. The Intermediate Nuclear Forces Treaty, the first agreement to eliminate an entire class of American and Soviet nuclear weapons, was signed by American President Reagan and Soviet General Secretary Gorbachev in Washington on December 8, 1987. The signing was the culmination of six years of bilateral negotiations on medium-range missiles. (Editor)

Ambassador Douglas Roche

CHAPTER 15
Canadian Circumpolar Relations

AMBASSADOR DOUGLAS ROCHE

How fast is the world changing today? I will tell you how fast the world is changing. In September the University of Southern California and the University of Illinois are going to play a football game that will be televised around the world and that game will be played in Dynamo Stadium in Moscow. (Naturally it is being called the *Glasnost* Bowl.)

All of us can think of illustrations that reflect what an extraordinary moment we are passing through in the world today. The Cold War that has dominated and poisoned relations between East and West since the end of the Second World War is ending just as a recognition is taking hold that global problems of the environment, development and staggering

Douglas Roche is a visiting professor at the University of Alberta. At the time of the inquiry, he was Canada's ambassador for disarmament and chairman of the United Nations Disarmament Committee. He served in Parliament from 1972 to 1984, where he specialized in the fields of development and disarmament. He is the author of eight books, the most recent being United Nations: Divided World *(Toronto: NC Press, 1984), an examination of the United Nations in the context of contemporary issues. Mr. Roche has served as president of the United Nations Association in Canada, international president of Parliamentarians for World Order and honorary president of the World Federation of United Nations Associations. He holds honorary doctorates from St. Stephen's College, Edmonton, Simon Fraser University and the University of Alberta, as well as the Christian Culture Gold Medal Award, the Alberta Premier's Award for Excellence and the Peace Award of the World Federalists of Canada.*

debt can be solved only by new international partnerships. This turning point in the human journey has created new hopes for peace in people everywhere.

Out of everything that we have learned in the decade of the 1980s—a decade of grave human suffering sharply contrasted with new processes of enlightenment—one overarching fact stands out: peace is a multi-agenda process; it involves economic and social development as well as arms control measures, and the protection of human rights as well as environmental security. The agenda for the twenty-first century, which is now only 129 months away, is already claiming our full attention: the danger of nuclear annihilation; regional wars using conventional weapons; the gap between North and South; the danger of overpopulation; the despoliation of the environment. Although all these problems are enormous, we have acquired the power to protect and sustain life. But to sustain life in its many splendrous forms on this planet, we need a bridge to a future of collective security, and the name of that bridge is international cooperation. It is cooperation, not confrontation, that will bring us peace inseparable from sustainable living.

It is with that theme of international cooperation in mind that I have come here this morning to address the topic of Canada's approach to the Arctic. My point can be briefly stated: the Arctic must be an arena for international cooperation.

Canadian foreign policy has traditionally been grounded in the firm belief that military confrontation can be forestalled or prevented by civilian cooperation. That is why Canada is a strong proponent of the economic and social activities of the United Nations system, as well as ancillary peacekeeping, development assistance, environmental protection and a host of other international cooperative programs. We are the thirtieth-largest country in population, and yet we are the fourth-largest overall contributor to the vast range of United Nations work. We support these programs for humanitarian reasons to be sure, but we also believe that they are a necessary complement to arms control, disarmament and other direct efforts to reduce military tension.

As essential as it is to strive to control and reduce the military hardware that threatens us, we cannot develop the sense of affinity and mutual dependence that will guarantee lasting international harmony unless we invest heavily in civilian cooperation at the same time. I am emphasizing this close connection between peace and active cooperation because this connection is the basis of the Canadian government's approach to international cooperation in the Arctic. We Canadians are well known for our special emotional attachment to the Arctic. We worry, and properly so, about any threat to its peoples, its wildlife and its environment. I think it was Franklyn Griffiths who cautioned us about what he terms "the Arctic sublime... a mysterious, poetic place to be understood with the heart rather than the head." I think that what he is telling us is not to form assumptions too hastily about what will and will not be good for the North. Perhaps we ought to get more northerners themselves to tell us—we started to see some of this yesterday—what kind of cooperative activities among the circumpolar countries will develop the trust that will develop global peace and security. In fact, the process of expanding and broadening the existing tenuous base was begun by the peoples of the Arctic themselves.

The age of easy accessibility to the Arctic is only about two or three decades old. Until the late 1970s international cooperation in the Arctic consisted primarily of sporadic scientific cooperation and occasional agreements like the 1973 polar bear agreement among the Arctic Rim states. In 1980 the Inuit of Canada, Alaska and Greenland formed the Inuit Circumpolar Conference (ICC), which has led to ongoing cooperation on a wide range of cultural, educational, economic, social and political matters. I think we owe a great debt to that organization. The ICC and similar groups, such as Indigenous Survival International (ISI), have inspired many Canadians and have opened our eyes to the breadth of the potential for international circumpolar cooperation. There is an important lesson here for all of us in this story of indigenous cooperation.

Whatever we do and whatever initiatives we take, we must measure our activities in terms of the direct benefits that flow therefrom to northern Canadians. Cooperation must be related to the national goals of the circumpolar countries, and one of Canada's primary objectives is to protect the Arctic and its resources and to offer a just and secure future to our northern citizens.

The government assists northerners in their efforts to build circumpolar links in several ways. In addition to providing financial and other direct support to the ICC and ISI, Canada has urged other countries to support and participate in these organizations. We are particularly pleased that the Soviet Union now appears ready to respond positively to our urgings to allow Soviet Inuit to participate in the ICC's general meeting in Greenland this summer.

The government has taken measures to increase contacts between Canadians and Greenlanders, such as opening an honorary consulate in Greenland and encouraging interested provincial and territorial governments to explore cooperation with Greenland's home rule government. In addition to direct northern input into policy, the Canadian government understands that the circumpolar North offers opportunities for international cooperation that we are just beginning to tap.

As you know, in 1985 the Special Senate and House of Commons Committee on Canada's International Relations recommended a distinct northern dimension to Canada's foreign policy. The government agreed and has since been pursuing bilateral and multilateral circumpolar cooperation in a variety of ways. Happily, other circumpolar countries and the circumpolar peoples appear equally interested in forging these new Arctic links. The result is that today we are witnessing more contact, communication and cooperation than we have ever seen. This process is just beginning. There are long roads ahead, but I believe that this intensely human process is conducive to long-term peace and security.

It should be noted that the government has difficulty with the proposition that the Arctic countries should get together and devise a special regime of restricted military activity. This view has nothing to do with the desire to increase military activity in the North. In fact, Canada's military participation in the North is very limited. Our whole defence expenditure in the North is less than one-tenth of one percent of total defence expenditures, and only one-quarter of one percent of Canada's military personnel are stationed in the North.

In a new era in which East and West are making progress in global achievements for peace and security, a more comprehensive approach to peace may be to keep the Arctic, or any other region, within that global process. The government is committed to peace and security in the Arctic in the same way that we are committed to peace and security the world over. Does it not follow then that discussions on peace in the Arctic are encompassed as part of discussions on global peace and security? I think of the Stockholm and Vienna forums as illustrations. Is it not those discussions that can and must reduce the level of weaponry in the whole world, including regions like the Arctic that are so precious for the preservation of the environment and native peoples' values?

Of course, there are national—or rather state-to-state—aspects to international relations in the Arctic, and Canada is pursuing these with equal enthusiasm. For many years we have attempted to engage the Soviet Union in cooperation in the Arctic. Before 1984 the Soviets were not ready to consider cooperation outside the hard sciences, while Canada insisted on including social and cultural components. As we know, in that year Canada and the Soviet Union signed a protocol on cooperation in the Arctic dealing largely with scientific and educational exchanges. The protocol has resulted in increasingly frequent and fruitful visits of Canadians and Soviets to each other's Norths and indeed has proved so successful that last fall the two countries initialled an expanded agreement that will broaden the scope of our cooperation. It is a real circumpolar success story and part of the whole aspect of global change today. I think this contributes directly to furthering peace and security between our two countries.

I might add that our agreement with the Soviet Union reflects Canada's long-standing insistence that social, cultural and educational questions of interest to northerners be included. Northern Canadians participate in a major way in these exchanges, and their provincial and territorial governments are major financial supporters.

Cooperation and exchanges between Canada and the Nordic countries and Alaska are also increasing. Because of our friendly relations with these countries there is less need for more formal agreements between national governments. Most of this activity is now occurring naturally; it is part of the daily activities of a growing number of private and public bodies interested in the North, such as universities, northern institutions, research institutes and of course The True North Strong and Free Inquiry Society, as manifested by this inquiry.

Only two years ago President Gorbachev, in an important change of policy, signalled a new desire of the Soviet Union to engage in multilateral cooperation in the Arctic. Since then, and we heard a good deal about this yesterday, the eight Arctic countries have tentatively agreed to the creation of a nongovernmental international Arctic science committee and are now considering, as we heard, the Finnish proposal to work together to address environmental protection in the Arctic. Last month Canadian parliamentarians met with their circumpolar counterparts in Moscow. Also last month Canada invited scientists from the Arctic nations to attend a meeting of Canadian scientists to consider the new disturbing evidence of toxic contamination in the Arctic food chain. This sudden almost meteoric rise in circumpolar cooperation is one that Canada supports and encourages heartily.

Finally, Canada has succeeded in reducing potential tensions in the Arctic through certain specific international legal instruments. Many years of hard arguing by Canada led to special environmental protection for the Arctic and other ice-covered areas in the Law of the Sea Treaty. Our 1987 agreement with the United States on cooperation in the Arctic did not settle our differences over the Northwest Passage completely, but it went a long way toward recognizing Canada's right to control surface shipping in those waters. The spectre of ecological damage through oil spills in eastern Arctic waters is mitigated by the 1983 Marine Environmental Cooperation Agreement between Canada and Denmark.

I think that is about as much as I want to put out here in these opening comments. You can see that a great deal has happened in the international Arctic in this decade. It didn't all flow from any one grand design but was built brick by brick by people who cared about the North.

I do not want to be accused of being overly optimistic about the Arctic. Let me just give one or two cautionary notes. Despite all this encouraging progress, the Arctic remains a sensitive area in many aspects. There are numerous boundary disputes. Some circumpolar countries have seen their

Norths ravaged by war in living memory, while others have economic lifelines in the Arctic about which they are understandably sensitive. There are many similarities within the circumpolar world but a lot of differences too. I think that international Arctic cooperation will not be built on well-intentioned proposals by themselves. Rather it will be built on the cautious, informed, persistent view that an enlarged cooperative spirit among the Arctic nations will make a profound contribution to global peace and security. The Arctic is very much a part of the multi-agenda process for peace.

CHAPTER 16
Canadian Defence Policies and Activities in the Arctic

MAJOR-GENERAL DAVID HUDDLESTON

It is a pleasure for myself and my colleagues to come back to this forum— not the same individuals but representing the same organization—and it is one that we wouldn't miss. I mention my colleagues, because I'm not capable of answering all the questions you may have; I came with a few friends who represent the various elements of the Canadian Forces. Indeed, our sailor is a submariner, so if you have your submarine questions ready at the coffee break, look for the gentleman in the dark blue suit with four stripes. We also have a couple of airmen, one down from Yellowknife, who is very familiar with all of the activities of the Canadian Forces north of 60, and one who may not admit it instantly but professes to be quite an expert on cruise missiles. We also have a soldier who spent some time in the Middle East. If you would like to talk about peacekeeping, please take him on. And don't let me off the hook either.

On the topic of communication, I should say that I regret that you got a rather unfortunate impression yesterday of our new minister. I have travelled with her frequently in the course of her first few weeks in our department. I know of nobody who is more willing to address the controversial issues of her portfolio than Mary Collins. Indeed she spent all day Friday in Vancouver speaking to disarmament groups there at their request, and I am sure that, if you are near her constituency in Vancouver, and wish to meet with her to discuss your concerns, she will be very willing to do that.

My topic this morning is simply a description of the activities of the Canadian Forces in the Arctic over time and how we see some of the strategic issues that concern that region.

Let me begin by asserting that the security of the Canadian Arctic is in-

Major-General David Huddleston

tegral to the security of Canada as a whole. There is therefore good reason for military activities there that are entirely defensive in character. They support both national and alliance defence objectives. Many of these activities have contributed directly to the development of the North, and many continue to do so but receive little publicity. Let me talk for a few moments about what the Canadian Forces have done in the Arctic, what they do now and why.

The history of the Canadian Arctic from the nineteenth century on is a history replete with military ventures. These were aimed initially at exploration and latterly at development and defence. The search for the Northwest Passage was in part strategically inspired. We are all aware of the exploits of McClure and Franklin, but we are less aware of the contributions to Arctic development made by the fledgling Canadian air force in the 1920s and 1930s—when General Greenaway was a young recruit, I think—when bush flying, ice patrols and communications in the Arctic were pioneered with quite primitive flying machines. During the Second World War, staging routes across the Arctic were established for the ferrying of aircraft both to the Soviet Union in the west and to the Allies in Europe to the east. These requirements resulted in the construction of airfields, communication systems, roads and pipelines. Northern air operations and the need for better weather forecasting led to the establishment of weather stations in the North by both the Allies and the Germans.

Direct military interest in the Canadian Arctic was a result of the development of intercontinental bombers capable of delivering nuclear weapons across the North Pole. However, the Soviets had, since the end of the Second World War, continued to develop their naval facilities on the Kola Peninsula and to pioneer the northern supply routes between Murmansk and Vladivostok. With the emergence of the intercontinental bomber, the strategic depth provided by the Canadian Arctic became a central element in the protection of North America.

In the mid-1950s, construction was started on the Distant Early Warning (DEW) Line. The DEW Line ran some fifty-eight hundred kilometres along the top of mainland North America and across Baffin Island and

Major-General David Huddleston is Associate Assistant Deputy Minister (Policy) in Canada's Department of National Defence. He holds a law degree from the University of Glasgow. He joined the RCAF on his arrival in Canada and subsequently undertook assignments in France and Germany. After completing National Defence Staff College he was appointed commanding officer at Cold Lake, Alberta. He served at NATO headquarters in Brussels and at Fighter Group headquarters at North Bay, Ontario, before promotion to Commander 1 Canadian Air Group. His awards include the Canadian Forces Decoration.

177

Greenland. The primary mission of the DEW Line was detection, but it had a very important secondary mission as a communications link between Norway, Iceland and North America. The DEW Line was a unique form of Arctic militarization in that its purely military function—to provide early warning—was performed largely by civilians and was not reflected in significant numbers of military personnel stationed in the high Arctic.

In the immediate postwar years, the Royal Canadian Air Force began the long and important national task of surveying Canada from the air and for the first time providing high-level photographic imagery across the high Arctic. Naval interest in this area led in the mid-1950s to the commissioning of Canada's first Arctic patrol vessel, HMCS *Labrador*. Some of you will remember her famous transit of the Northwest Passage in September 1957. The 1960s saw the establishment of new weather stations in the Queen Elizabeth Islands, a network of low frequency navigational aids in the Loran Chain, a pipeline from Haines through the Yukon to Fairbanks, Alaska, dispersal bases for aircraft and communications for the Ballistic Missile Early Warning System. These defence projects provided airfields, roads, communications, navigational aids and other infrastructure of a higher standard and at earlier dates than would have been possible if only civilian development of the Arctic had occurred.

The importance and relevance of the DEW Line declined during the years when the principal transpolar threat to North America was that of intercontinental ballistic missiles, whether ground- or submarine-launched. The advent of nuclear propulsion for submarines permitted under-ice operations, and the increasing range of submarine-launched ballistic missiles significantly lessened the requirement for these launch platforms to venture far from their home ports. The newer classes of Soviet ballistic-missile submarines, when stationed in the Kola Peninsula, can now seek sanctuary in the Barents Sea or under the polar ice while remaining within range of their North American targets.

The revival of the manned bomber and the development of the unmanned cruise missile that can be air-launched have renewed the strategic significance of the Arctic as an area where early warning of an attack is possible. The cruise missile can also be launched from submarines at the ice edge or in polynyas,[1] or from surface ships in open water. The obsolete DEW Line has given way to the North Warning System in response to these developments. Unlike the DEW Line, the Canadian section of the North Warning System is operated and supported by Canadians, and its radar picture is fed directly into the Canadian Region Operation Centre of the North American Aerospace Defence Command (NORAD) where incidentally the Department of Transport has a seat and for the first time can watch those commercial airliners passing through northern Canada and talk to them. So not all is quite as bad as General Greenaway painted it.

The importance of this new system has been heightened by the increasing emphasis being placed on cruise missiles by both the United States and the Soviet Union. While cruise missiles will be limited in number by strategic arms reductions, their relative significance will increase if the Strategic Arms Reduction Talks (START) are successful in achieving deep cuts in the intercontinental ballistic missile inventories of both the Soviet Union and the United States.

As part of North American air defence modernization efforts, existing northern airfields will be refurbished as forward operating locations for our fighter aircraft, so as to permit them to intercept penetrating bombers before the bombers release their cruise missiles.

The defence White Paper of 1987 calls for the establishment of an Arctic training centre, which will be located at Nanisivik, to train Canadian Forces personnel and units to operate in Arctic conditions. To protect our Arctic waters, the White Paper also calls for a fleet of nuclear-powered submarines, which would have the capability to patrol Canadian internal and territorial waters in the Arctic, although their primary operating area would be in the Atlantic and Pacific.[2]

To support the great variety of ongoing Canadian Forces activities in the North, the Canadian Forces Northern Region Headquarters was formed in May 1970 in Yellowknife. This headquarters maintains close contact with the operational commands and serves as a link between them and the northern settlements in which they operate and exercise. It is also the agency through which National Defence Headquarters coordinates assistance to, and communications with, federal and territorial government agencies north of the 60th parallel.

The Northern Region encompasses the land and waters of Yukon and the Northwest Territories, which cover 3.9 million square miles—more than one-third of Canada. The Northern Region Headquarters controls an air transport and rescue detachment based in Yellowknife, which flies approximately two thousand hours a year on skis, floats and wheels. The aircraft engage in reconnaissance, supply, liaison visits, search and rescue, and emergency air evacuation. We have communication squadrons at Yellowknife and at Whitehorse. The Regional Headquarters also controls the northern group of Canadian Rangers and officers of the Cadet Instructors List.

The northern group of Rangers, formed in 1947, is a unique component of the reserve forces numbering about eight hundred Dene, Métis, Inuit and whites divided into forty patrols, eighty-seven percent of whom are Inuit. Their primary roles are to report suspicious or unusual activities and to collect local information to assist Canadian Forces operations. In their distinctive red hats and arm bands, they are a visible manifestation of Canada's sovereignty in the North. The regular Canadian Forces cannot practically or economically provide such a ubiquitous presence.

Plans call for a gradual expansion of the Rangers in the North to about one thousand Rangers in fifty patrols. The Rangers perform a host of duties, acting as guides and survival instructors, instructing in native skills, participating in search and rescue, assisting in local defence and assisting the RCMP or local police in the discovery and apprehension of hostile persons. It would be difficult for the Canadian Forces to operate in the far North without the help of Rangers, who can be justifiably proud of their contribution to the defence of Canada.

The cadet movement in the North continues to be a most important training ground for citizenship and character. In many communities the local cadet unit is the only organized youth activity. There are currently some seven hundred cadets enrolled in fourteen army cadet corps and five air cadet squadrons in the North.

Sovereignty activities in the North comprise regular visits by staff of the Regional Headquarters to communities and regular force exercises in isolated areas. Northern patrols, or NorPats as we call them, are flown over the region's airspace by Aurora aircraft from Comox, British Columbia, and Greenwood, Nova Scotia. These flights provide regular reconnaissance of Canada's Arctic and are also a source of aerial photography, adding to our knowledge of the area. They assist in monitoring fisheries activities, environmental concerns and wildlife movement.

The collation of information acquired by ships and aircraft on northern deployments is one of the Northern Region Headquarters staff's principal tasks. This information takes the form of maps, charts, site plans, photographs and statistics on everything from airfields to oil rig locations. The mapping and charting establishment of the Canadian Forces conducts annual field survey operations in Canada's North in support of topographic mapping projects. The most recently completed project involved surveys of Ellesmere Island.

I think the Canadian Forces are good corporate citizens when it comes to the protection of the environment. All new activities are based on four general principles: first, to obtain an objective environmental impact assessment; second, to minimize disruption to local people; third, to consult with local people; and fourth, to the extent possible, to direct economic benefits to the local community. All new projects are subject to an environmental assessment and review process. When significant impact on the environment is anticipated, an independent assessment is made and public input is invited through hearings. Recommendations are offered to both the Minister of the Environment and the Minister of National Defence. This process ensures that an objective third party is seen to have studied the impact of the project on the ecology, the economy and the social fabric of the affected locality.

A good example—it is somewhat controversial, but I think it is a good example of how far the Department of National Defence is prepared to go in order to minimize the disruption of people and societies affected by military operations—is the arrangement at Goose Bay. The Innu hunting parties are requested to advise the base of their planned movements, and the helicopters and light planes now regularly used by Innu hunters are tracked so that military aircraft can avoid overflying their camps and can be diverted around known locations of hunters and caribou herds.

An example of the extent of consultation by the Canadian Forces with local authorities is that now taking place with the Baffin Region Inuit Association, the Lancaster Sound Regional Land Use Planning Commission and affected communities with respect to the proposed establishment of a northern training centre at Nanisivik at the north end of Baffin Island. The training area associated with the centre is close to the Inuit community of Arctic Bay.

The fourth principle, that of directing economic benefits, to the extent possible, to local communities, involves offering first refusal on jobs and an opportunity to bid on any contracts to those communities and individuals directly affected. As you are aware, directed economic benefits are sometimes very difficult to guarantee in a competitive society such as ours, but they have worked well in the Western Arctic with the North Warning System. There negotiations are taking place between the Inuvialuit Regional Council and the Department of National Defence to encourage local businesses to bid on various construction contracts for the short-range radar sites located in their region. The department has also initiated a process to assist native business people to develop the skills required for the technical maintenance of radar equipment planned for these sites. Communities will be encouraged to contribute to the security of the unmanned radar sites through voluntary surveillance.

In short, our military presence in the Arctic is now, and will continue to be, defensive in nature and sensitive to environmental concerns. The naval, air and military activities that we undertake in the region threaten no one. We have fewer than four hundred members of the Canadian Forces stationed north of the Arctic Circle and we have no plans to base submarines, ships or major air units permanently in the far North.

By contrast, as referred to over and over again, the Soviet Union maintains a large Arctic-based fleet of 76 major war vessels, 171 submarines and 446 naval aircraft, for a total of over one hundred thousand naval personnel. In addition, two motorized infantry divisions with helicopter support and an extensive tactical air force are stationed in the Kola Peninsula. It is somewhat ironic that President Gorbachev, when making proposals for northern demilitarization, should have spoken at Murmansk, the centre of the only significant and continuing concentration of naval, military and air forces north of the Arctic Circle. I recognize of course that the Kola Peninsula provides the Soviet Union with the only reasonable base for its Atlantic fleet and that is a geographical fact one cannot dispute, but I do not particularly see the reason for having the extensive land and air forces there, which could serve none other than a threatening purpose.

These particular forces, however, are excluded from Mr. Gorbachev's specific proposals. Naval activity would be restricted in a large area of vital security importance to the North Atlantic Treaty Organization (NATO), namely the North Sea, the Baltic Sea, and the Norwegian and Greenland Seas. But the Barents Sea, which is vital to the Soviet Union, is excluded from such restriction.

The primary, although not the only, direct military threat to Canada is from nuclear weapons. Declaring the Arctic a nuclear-weapons-free zone or restricting certain naval movements as confidence-building measures will do nothing to reduce the threat from nuclear weapons, which can be moved from region to region. Indeed it could divert attention from serious arms control negotiations. The threat can only be reduced and finally eliminated through measures aimed at resolving the underlying problems in East-West relations. As was pointed out in a recent brief published at the Canadian Centre for Arms Control and Disarmament, "The sentiment underlying support for a Canadian nuclear-weapons-free zone would thus be better turned to a more realizable objective."

On March 6, I was privileged to be present with the minister, Mary Collins, at ceremonies to initiate the new negotiations between NATO countries and those of the Warsaw Pact on conventional armed forces in Europe. These negotiations, by tackling the basic problem of stability in Europe, will set the tone of East-West relations for the next decade. The Soviet Union, by Mr. Gorbachev's own admission, enjoys a significant advantage in conventional land and air forces stationed in Europe, and until such time as verifiable, asymmetric reductions can be negotiated, it would be imprudent for us to inhibit the mobility of our maritime forces.

With respect to stability and the balance of strategic nuclear weapons, based on the renewal of contact between Mr. Baker and Mr. Shevardnadze in Vienna, we can look forward to further progress in the START talks this spring. These aim at a fifty percent reduction in the arsenals of the United States and the Soviet Union.

The negotiation of verifiable, mutually beneficial arms control agreements that lower the level of armaments while preserving the security of both sides will improve stability in East-West relations. It is through these agreements rather than through regional measures that the cause of peace will be promoted—a peace more evidently durable than that which Canadians have enjoyed these forty years.

Notes

1. Polynyas are persistent or recurrent areas of open water in fast ice or ice-choked seas. (Editor)

2. Canada's defence White Paper of June 1987 reaffirmed the government's solidarity with the North Atlantic Treaty Organization (NATO) and committed the country to a major military procurement program, including the commissioning of a fleet of ten to twelve nuclear-powered attack submarines. Less than two years later, in the April 1989 budget statement, the government announced that the submarine procurement had been cancelled as an economy measure. (Editor)

International Relations: A Government of the Northwest Territories View

Dennis Patterson

CHAPTER 17
International Relations: A Government of the Northwest Territories View

DENNIS PATTERSON

Qujannamiit.[1] Thank you very much. *Ullaqut.* Good morning. *Tungasuttitaugama.* I am being made to feel very welcome here. On behalf of the people and the government of the Northwest Territories, I am happy to have been invited by the organizers of the True North Strong and Free Inquiry Society to provide our perspective on international polar cooperation. This comes, if I may humbly say so, from the largest semi-autonomous jurisdiction in the circumpolar world. I refer to our area, not our population.

I remember a popular movie a number of years ago in which the late Peter Finch played a television prophet who shouted at his audience, "I'm mad as hell and I'm not going to take it anymore." Now many people in the Northwest Territories are starting to have the same kind of thoughts.

Dennis Patterson is government leader of the government of the Northwest Territories. He is additionally responsible for the Intergovernmental Affairs, N.W.T. Science Institute, Office of Devolution and Audit Bureau portfolios. He holds a B.A. from the University of Alberta and a law degree from Dalhousie University. He was admitted to the bar in Nova Scotia and British Columbia and worked with law firms in both these provinces before moving to Iqaluit in 1975. There he became the first executive director of the legal services centre, Maliiganik Tukisiiniakvik, which employs Inuit paralegal staff and is run by a board of directors. Mr. Patterson has represented his riding since 1979 and has held Education, Information, Status of Women and Aboriginal Rights and Constitutional Development portfolios, as well as chairmanship of key executive council committees.

International Relations: A Government of the Northwest Territories View

In the early years, chicken pox, measles, flu and other European diseases imported to the Arctic by explorers just about decimated our northern population. Later it was starvation. Today it is AIDS, Arctic contaminants, and trying to cope with the anxieties created by advanced cruise missile testing, low-level military exercises, the possibility of nuclear-powered submarines and the European boycott of trapped furs.

The "true North strong and free"? People in the Northwest Territories would like to think so. But maybe that isn't the case any more. Let me ask you a few questions.

What if levels of airborne toxic chemicals carried to the Arctic from other areas of the Northern Hemisphere increase to the point where contamination of the northern food chain forces our people to abandon their traditional dependence on "country foods"? How will you feel if the cultural lifestyle that dates back a thousand or more years is destroyed by the antitrapping movement? What if the guidance system on a cruise missile breaks down shortly after the missile is launched over the Beaufort Sea and it crashes on Inuvik's main street?

What if the United States had asked to test the cruise missile along a corridor originating in northern Ontario and terminating a couple of hundred miles north of Ottawa? Would the decision to allow the tests have been the same?

What if you don't have freedom from anxieties? Is it still possible to have global peace and security?

Has anyone out there got the answers? Can anyone relieve our anxieties in the Arctic by explaining the probabilities, the technicalities and the scientific theories to a population that speaks eight different languages, none of which has the appropriate word equivalencies?

In our rush to succeed we have walked on the moon but haven't paid enough attention to our ecosystems, to the world's environment or to the peace and security of mankind.

Success in aviation brought the twentieth century to the Northwest Territories. It is going to take a global effort and the hard lobbying of concerned people like ourselves to make sure it is habitable for future generations.

To many people, the Canadian Arctic has always symbolized all that is good from an environmental perspective: clean water; clean air; majestic, pristine landscapes; an abundance of wildlife resources and fish stocks; and a richness in undeveloped mineral and petroleum resources. Unfortunately however, there is a tendency to overlook the people who live there—to forget that we have strong opinions and beliefs and an abundant capability of managing our own affairs.

For the record then, the Northwest Territories is the homeland for a native majority that considers the land a sacred trust and where any developments must take serious consideration of that reality. Of equal importance is that it is a place of residence for people who are determined to exercise control over their land. It is home to people who want to decide their own political and constitutional futures, rather than have them shaped by someone else in a distant national capital or by other premiers and the prime minister at first ministers' conferences, through such instruments as the Meech Lake Accord, which definitely left us and our Yukon neighbours out of the process, although together we cover forty percent of the land mass of Canada.

Unlike southern jurisdictions, the Northwest Territories does not have a government based on the political party system. Decisions by the legislature are reached by consensus, and we are proud of that. That is the northern way, a process that I believe is a direct reflection of aboriginal values in the areas of decision-making and conflict resolution.

Because of this difference in systems, however, it is sometimes hard for us to win respect and to get our points across, despite the fact that we are a bona fide government, fully elected and, I believe, representative of the people we serve.

I know there were calls here yesterday for northern self-government, but I want you to know that, while there is much yet to be done, I consider our government to be well on the way towards northern self-government. Half of my cabinet of eight are aboriginal people, including two women. A full three-quarters of us, like myself, represent constituencies with aboriginal majorities. Furthermore, over two-thirds of our legislature—which operates simultaneously in five aboriginal languages—are aboriginal people.

In Nunavut, Inuit communities such as Arviat (formerly Eskimo Point), whose young people you heard speak so eloquently yesterday, the local government, health and education services and others are run entirely by Inuit. After land claims are settled, the Inuit control over lands and resources will be even stronger. And yet it seems easier for some southern-based northern experts to attract national attention than it is for resident northern leaders speaking from within the North—and their views are not necessarily always the same, let me tell you. For that reason, I would like to thank Ambassador Roche for suggesting that it is time for northern residents to be given a chance to participate and to speak out. It is the reason why I was so anxious to accept this invitation to speak with you today.

I was asked to talk to you about international cooperation in the Arctic, an area in which the Inuit Circumpolar Conference—and its Canadian president, Mary Simon—have done so much important work recently.

Never before has the need for such cooperation been so pressing. Never before has the need to learn and to develop global solutions to Arctic problems been so great.

The record of the government of the Northwest Territories in circumpolar relations dates back to the early 1970s, when the head of the government, Stuart Hodgson, was appointed by Ottawa. I can recall when he and a number of his officials would fly from Yellowknife to Greenland to visit counterparts in that country. He also travelled to the Soviet North in 1971 with Jean Chrétien, the Minister of Indian Affairs and Northern Development. That was the start of what became a series of formal and informal exchanges between officials, local inhabitants and business people. All were encouraged to develop formal and informal relationships and to learn through open and frank exchanges of information.

Since that time, the process has matured, and in recent years it has broadened to include exchanges with Alaska and more intensive exchanges with the Soviet Union.

These exchanges have demonstrated to us that the people of circumpolar countries share common problems, be they social, economic, cultural, scientific or technical, and that on many occasions there is a far greater commonality of understanding between ourselves than between the Northwest Territories and our national central agencies and southern jurisdictions. As a matter of fact, a memorandum of understanding will be signed by Greenland Premier Motzfeldt and myself during a visit there by members of my cabinet and representatives from the legislature from April 10 to 14, 1989.

In any case, it has been over the last ten years particularly that the government of the Northwest Territories has played what we believe to be a significant role—at what you would consider perhaps the provincial level—in international polar relations. One of the first concrete examples of cooperation involved the small hamlet of Sachs Harbour on Banks Island. In 1974 that community agreed to provide musk-oxen to reestablish a population that had vanished a thousand years earlier in the Soviet Union. I think it is very significant that this good-neighbour policy of a small Inuit community led to the reintroduction of musk-oxen to the Soviet North. Our Deputy Minister of Renewable Resources went to look at those same musk-oxen last summer and tells me they are not only surviving but thriving.

Ambassador Roche mentioned the series of Soviet exchanges we have been very involved in that are governed under a protocol signed by the Soviet Union and Canada in 1984. The protocol covers four major themes: geoscience and Arctic petroleum; northern environment; northern construction; and ethnography and education. Two of these are chaired by deputy ministers of the Northwest Territories government; I am happy that our government is taking such a large role in the Arctic science exchange program on behalf of Canada.

The exchanges have resulted in a number of interesting ventures, including the design of joint Northwest Territories–Soviet projects for a settlement in the Autonomous Soviet Socialist Republic of Komi and a child care centre designed by the Soviets for Tuktoyaktuk. As well, they have resulted in the sharing of scientific and technical information on northern construction techniques, education, law, the environment and wildlife resources.

There is an important human side to this too. Last year a group of Canadian social scientists made up of native and non-native professionals visited Chukotka in the Soviet Far East. During a visit to a boarding school, a young Siberian Eskimosa (as they are called) had difficulties in explaining her experiences in the boarding school. She was embraced by the two Inuit members of our delegation, both of whom had attended a boarding school in Canada and immediately understood her difficulties, resulting in a bond of mutual understanding, trust and some tears of joy.

Another example, which I personally experienced, was during my visit to the Republic of Yakutskaya, an autonomous republic in Siberia—an area slightly smaller in size than the Northwest Territories. The chairman of this region explained eloquently to us his difficulties with financing, centralized control and the insensitivities of central agencies, all of which was not too dissimilar to some of our own experience in northern Canada with distant federal authorities in Ottawa.

On a personal note, I want to tell you that, as a member of that delegation and now happily having received many Soviet visitors to the Northwest Territories, I have been overwhelmed by the warmth and openness of our Soviet circumpolar neighbours. Everywhere we went everyone we met, from government officials to ordinary citizens and students, talked movingly and sincerely about their desire for friendship, cooperation and peace between our two great northern countries. And while we talked about education and cultural exchange and permafrost construction and social issues, we all felt and knew that we were doing our part on a human level towards peace between peoples who have much more in common than they have differences. Our ability to discuss and understand each other's problems has reinforced our belief that there is a natural bonding agent among circumpolar northerners that goes far beyond international borders and politics.

This thought leads me to conclude that the time has come for northerners to make our voices heard in a manner that will prompt national and international political leadership to respond to the issues of peace and security in the Arctic. This was referred to by Soviet leader Mikhail Gorbachev in Murmansk in October 1987 when he proposed a conference of Subarctic states to coordinate research in the Arctic. Included would be cooperative arrangements in polar, scientific, environmental and economic endeavours with Canada and other Western countries. At the same time he proposed that the Warsaw Pact and the North Atlantic Treaty Organization (NATO) begin consultations on scaling down militarization and restricting naval and air force activity in the Baltic and the Greenland, Norwegian and North Seas. Such action, he said, could lead to demilitarizing the top of the world and could turn the Arctic into a zone of peace.

Gorbachev made mention of cruise missile testing in Canada and the possible acquisition of nuclear submarines. However, as General Huddleston has pointed out, Gorbachev made no mention in his remarks—and, I think, cleverly—of the Soviet Union's Kola Peninsula, one of the most heavily militarized areas of the world.

The point I want to make though is that, despite the rhetoric, we have been presented with a starting point—a point from which an Arctic peace zone may some day be a reality—and I think we have already made some progress since that time. Such an arrangement I know would be welcomed by the legislative assembly of the Northwest Territories, which took steps a number of years ago to unanimously declare the Territories a nuclear-weapons-free zone.

Such a circumpolar zone of peace, I am sure, would also be welcomed by the Inuit, who are expressing concern over the contamination of their environment and over Canada's plan to purchase nuclear submarines to patrol beneath the Arctic ice. I want to say to you again how delighted we are with the strong initiatives towards circumpolar cooperation by the Inuit Circumpolar Conference and its very capable first Canadian president, Mary Simon.

I also know that such a plan for an Arctic zone of peace would be welcomed by the Dene of the Northwest Territories, many of whom deeply regret that uranium taken from their homeland was part of the payload in the nuclear bombs used on Japan to end the Second World War in 1945.

The problem surrounding this call for an Arctic peace zone, of course, is that the proposal came from Gorbachev instead of from a Western leader. As one reporter wrote at the time, "Officials are asking whether the Soviets are sincere; or is it a mischievous Kremlin trick to exploit Ottawa's dispute with Washington over sovereignty in the Northwest Passage in order to divide the countries of the West."

I am not here today to judge the veracity of the Soviet leader. I am here, however, to say categorically on behalf of our people that it is time Canada accepted this challenge to enter into meaningful and productive dialogue on matters of joint Arctic interest.

Dennis Patterson

My message is simple. Let us not let suspicion over Soviet seriousness about demilitarization block this opportunity. Start the talks at the highest level possible. The clock is ticking. The potential is great. The possibilities are endless. Thank you. *Qujannamiraaluk*. Thank you very much.

Notes

1. Each Inuktitut expression is followed by its English equivalent. (Editor)

CHAPTER 18
Question Session

JOHAN JØRGEN HOLST, MAJOR-GENERAL
DAVID HUDDLESTON, DENNIS PATTERSON,
AND AMBASSADOR DOUGLAS ROCHE

Gwynne Dyer
I have listened with considerable interest to three of the speakers we have just had, each in his own way telling us that nothing must be done about the Arctic until we have progress in conventional arms talks in Europe[1] or in the Strategic Arms Reduction Talks (START), and that what we do must somehow be linked to a global outbreak of arms control and disarmament measures. The Arctic mustn't be treated separately. It would be imprudent, it would be inadvisable, it would be difficult, to attempt to institute any kind of local nuclear-free zone, any kind of local arms reductions or regional arms control measures in the Arctic. I find it a bit hard to believe that that is absolutely, necessarily true or that we really must wait for progress in nuclear arms reduction talks, in START or the CFE talks in Vienna or whatever, or that we must wait for them to arrive at their conclusions before drawing our conclusions about the Arctic. In particular, I think this is urgent for Canada because, although, as Ambassador Roche said, we spend only one-tenth of one percent of our defence budget north of 60 at the moment, the primary justification for what the government admits to be an $8 billion program to buy nuclear submarines—$16 billion perhaps in some other people's estimation—is that these submarines must be able to operate in the Arctic. Somehow I think that's going to kick it a bit above one-tenth of one percent.

So I would like to make a proposal and I want to put it to these three gentlemen. I also intend to put it to the Soviet delegation when they come up here later today. I want to ask: What would be the problem with instituting a nuclear-free zone in the Arctic Ocean that would allow no nuclear-weapons-carrying vessels or aircraft (indeed no nuclear-powered vessels) to operate in the large area of the Arctic that impinges on our

193

shores, that impinges on the shores of anywhere except the Soviet Union? In other words, create a nuclear-free zone in the Arctic that would take account of the special conditions of the Soviet Union, which must base its Atlantic fleet in the Kola Peninsula, by saying they can't take anything nuclear more than five hundred miles from their own coast? Then allow for the problem of the Greenland–Iceland–United Kingdom (GIUK) gap—the only way the Soviets can get out to the Atlantic or indeed the only way the Americans can get into the Barents Sea to get at them—by declaring a zone within which, although the Soviets can steam out into the Atlantic if they want and the Americans can pay port calls in Norway, everybody has to advise in advance of their intention to pass through; everybody passes through on the surface—no submarines go through submerged; and nobody stays there or conducts exercises there. Such a zone could be between, let us say, at one end the Norwegian-Soviet border and those islands north of there, the Spitzbergen and Svalbard Islands, and at the other end somewhere down around the bottom end of Greenland, Iceland, the Faroes and somewhere along the Norwegian coast.

It seems to me that that would give the Soviets security within their necessary zone of operation and the necessary access to the Atlantic, and it would still permit the demilitarization, or at least the denuclearization, of the entire Arctic Ocean, save that area immediately off the Soviet coast where they require a sanctuary for their missile-firing submarines, pending progress in things like START. Could I put that to them, please?

Johan Jørgen Holst
First of all, I appreciate the sentiment that is expressed. But I think that it is necessary in order to be able to preserve the Arctic—and I think I am as committed to that proposition as is the questioner—to be very precise as to what is practical and what is doable. None of us, I think, said—I certainly didn't say—that we had to sit back and wait for things to happen elsewhere. I did warn that I don't think that measures specifically designed to regulate things only in the Arctic are a particularly helpful and fruitful approach at this juncture. But I think it is incumbent upon the countries most concerned about developments in the Arctic to take into account, to protect so to speak, the Arctic development interests in relation to these primary processes and negotiations that are taking place. That is no mean undertaking.

I think there are three difficulties about the proposition you just advanced. First of all, it does imply a sort of general territorialization of the open seas, and I would submit that that constitutes an enormous set of difficulties with respect to how you manage international order. How are you to ensure that other countries would not undertake the same thing, and where would we then be with respect to trying to develop a more enlightened international regime for the open seas? The second point has to do with verification. I know of no way in which to verify what you propose. I submit that things you undertake to do that cannot be verified tend to become destabilizing rather than stabilizing. They generate conflict rather than reduce conflict. My third point is that I don't see how you can distinguish between transit rights, which you acknowledged that, for instance, the Soviets would have, and presence rights. Why is it that countries that are particularly close to an area should have certain rights and other countries should not have the same rights to protection? You know my own country, for instance, borders on the Soviet Union. If we were to be excluded from association in military terms along with other protectors, why would that be a reasonable and legitimate order to introduce? I just don't buy it. I think this is not a workable proposition. I don't think it is the way we should work it. I think there are a lot of important things that can be done, and I think that I suggested some in my speech.

Douglas Roche

Mr. Dyer has made a proposal that I think on the surface could well be examined, but I must say that I would have some concern about it initially. Nuclear-weapons-free zones, where they work in the world—the South Pacific is one example—require that all parties, all countries within a region, fully participate, and it is not clear to me at the outset how a nuclear-weapons-free zone in the Arctic, which would be limited in Mr. Dyer's own definition of it, would act as an instrument of stabilization. In fact, I would enter an argument that it could be an element of destabilization.

However, there was something deeper in Mr. Dyer's question that concerns me. It is true that the panelists this morning have spoken about the need for an attitude of cooperation to be developed among the northern partners, certainly including the Soviet Union. I would think that ought not to be lessened, that is, the need for this new understanding that is coming into play among the Arctic countries as an element of, almost a precondition to, what might come in the future. I was quite intrigued yesterday by Professor Pharand's rather speculative views, and I thought it was an interesting line of thought that he was advancing as to where the non-military cooperation we are now seeing among the eight Arctic countries could lead in future days, given the whirlwind, in relative terms, of events taking place in the changing structure of East-West relations.

I think that it is only appropriate that I draw your attention here to the Consultative Group on Disarmament and Arms Control Affairs, composed of about fifty Canadians from coast to coast, which I chair. At a meeting on peace and security in the Arctic, which I convened about two weeks after the Murmansk speech of Mr. Gorbachev, the Consultative

Group made recommendations to the Government of Canada on ways for Canada to develop an Arctic policy that ensures Canadian sovereignty, protects the northern environment and people, and contributes to international peace and security.

And I think that that, along with the parliamentary committee, was an influence on what we saw in a rather formal exposition of Canadian policy yesterday in Mary Collins' speech. We are undergoing an evolution these days, which I think can have very positive gains in the future.

Lastly, the Consultative Group met about two or three weeks ago to discuss an arms control and disarmament agenda for the 1990s and produced a report that gives an indication of where the Canadian government can go in this quest for common security. I think that these reports of very concerned and informed Canadians who are advising the government today are material that would be of interest to the media in examination of the long-range prospects of Canadian security.

David Huddleston
I would simply add and reinforce my view that security is not synonymous with defence and arms control, that there are many other aspects to it, and that one should not have the sense that, while waiting for formal negotiated arms control agreements, one need do nothing. We listened to people yesterday talking about many aspects of cooperation in the Arctic that will contribute to a better understanding and to better East-West relations and that ultimately will complement those negotiated agreements and thus contribute to security. I do get the sense that there is a great deal of appeal in dealing with symptoms rather than causes, and the Arctic is a little piece of the puzzle. I would rather see us focus on the heart of the problem, which is the relationship that has existed for many years among states.

We in my department, as Ambassador Rodionov knows well, have in the last few months initiated many discussions with the Soviet Union on the sorts of contacts and the sorts of cooperative useful discussions that we can have. Our National Defence College will visit the Soviet Union for the first time in ten years this year. We are already arranging mutual port visits. These may seem to be somewhat superficial, but I don't believe they are superficial. They indicate a significant change of direction and a recognition of the fact that East-West relations are at a turning point. All of these things are in their own way small, but they are all actions, they are all positive and they contribute to the larger picture.

Ann Medina
I am following through with a question from yesterday. Mary Simon felt that the Canadian government was not really including the Inuit in many

policy-making decisions, although, Mr. Roche, you talked about the Arctic being the area for international cooperation. What steps towards cooperation with the Inuit have been taken? There are problems with the funding of the Inuit Circumpolar Conference, for example. How much has the government put into the funding to make that cooperation very real? Also I understand that, for the meeting of December 1988 that Canada had with the Soviets, it was the Soviets, not the Canadian government, who invited members from the Conference. Perhaps you could tell us where you think that you are not urging or following through with cooperation with the Conference and where you feel that you could take further steps.

Roche
First, I certainly want to identify with the thrust of your comment and question. For me, the most poignant moment of this conference was yesterday when two Inuit young people stood at that microphone and made what was really a cry from the heart for participation in the economic and political development of their society. I think it is clear that insufficient Inuit participation has taken place so far. In my speech I referred to the developments that have taken place, that is, the Circumpolar Conference, and so on. There has been some funding. I think that External Affairs is noting this inquiry very carefully and probably will take back to Ottawa a feeling—and I think a recommendation—that increased funding be made available for the relevant organizations to participate in their own development. I say that because for me personally the comments made yesterday were very telling indeed.

Linda Hughes
Our understanding is that the Inuit Circumpolar Conference in Canada gets a grant of $100,000 each year, and that it isn't regular funding—they have to fight for it every year—while in Greenland the funding given to their counterparts is $400,000 each year.

Roche
I am pretty sure all of this is going to be looked at in light of this inquiry.

Hughes
I wonder, to follow up with Mr. Patterson, if because of what has been said here about funding, you think that there is a level of hypocrisy here. It seems that throughout the inquiry many people—and most of them are the southern leaders that you are talking about—have talked about how important and central it is to have the Inuit involved. Yet there seems to be a credibility gap between what they say we should have and what they are actually doing at the federal level.

Question Session

Dennis Patterson
Perhaps I betrayed in my remarks some of the frustrations that northern people feel who are living in the North, including of course the Inuit residents who have established sovereignty in the North for Canada, in many cases after having been really forcibly relocated without much choice to places like Grise Fiord and Resolute Bay.[2] We also feel left out of the Canadian mainstream in the decision-making process. I guess as a mere territory we can't claim much authority in external affairs matters, and we are certainly not trying to usurp the national government's role to speak for Canada in international matters. But I quite sincerely believe that Canada's presence internationally with regard to circumpolar affairs and in areas like the contamination of the Arctic would be enormously strengthened and enhanced by more active participation of the people who live in the region, their governments and their organizations.

You asked me about a level of hypocrisy. Those are strong words. I think that there has been rather a lack of consideration, which I think is changing rapidly. On the fur issue, for example, the government of Canada changed its approach to the involvement of northern native people quite dramatically. At first there was a reluctance to consider the European animal welfare movement an important issue, but I must say that in recent years Mr. Clark in particular has become a very strong advocate. I think it was Canada's intervention with Britain that was quite critical in turning that fur-labelling issue around. So I think the frustration that we feel is diminishing. A lot has to be done, but I think progress is being made. I think eyes are being opened and our presence is being felt, and people like Mary Simon are carrying the message very strongly. So I am optimistic that northern peoples will be involved in discussions about the contaminant issues as part of the Canadian delegation. I think there have been serious problems, but I am optimistic that things are turning around and that our government is going to start speaking out more and more. As well, we are going to be a lot more aggressive and demanding and, we hope, more persuasive than we have been in the past. So I think things are going to improve.

Dan Heap (Member of Parliament, Trinity-Spadina)
I am also here as a member of Veterans Against Nuclear Arms. My question is to Mr. Patterson. Mr. Holst this morning was the first speaker at the inquiry to declare support for the principle of common security between the NATO bloc and the Warsaw Pact bloc. More and more Canadians support that principle, which means that if the Soviets don't feel safe, we aren't safe, and if we don't feel safe, the Soviets aren't safe. The Canadian government denies this in its White Paper on defence[3] by claiming that Canada is in danger of being invaded by the Soviet Union. Since

our agenda yesterday and today is loaded with representatives from the Department of National Defence and the Department of External Affairs, there is no time on the platform for parliamentary opposition. I would ask Mr. Patterson if he would tell us whether he believes that the Soviet Union really wishes to invade Canada or whether it is planning a first military strike against North America, and if so why he believes that and why he thinks they would do it. In other words, is the military budget and program of the Canadian government what the people of the Northwest Territories want?

Adrienne Clarkson
I would like Mr. Patterson to answer that, but I think it would also be interesting to have General Huddleston reply to it.

Patterson
Very briefly, my contacts with the government officials and the ordinary people of the Soviet Union indicate that there is indeed a desire to invade Canada, but it is a desire to invade Canada on a friendly basis and get to know us and get to be friends and brothers. I have to say about the White Paper, with which we have been very involved and concerned, that there are many good things about it. I want to be fair. We are pleased with the North Warning System. I think it is a defensive system. I think it will help civilian aviation, and I think it has been planned in a way that will address the genuine concern about providing some local employment opportunities and benefits, including improved airstrips and the like for northern people. We also are proud of the Rangers and the air cadets. There are a lot of good, I would say peaceful, military activities happening in the North that we are proud of, but our people are very concerned about the need for, and expenditure of money on, nuclear submarines. We would like to see money spent in the name of sovereignty in the North on things that we really need and can see like ports, roads, highways, airstrips and docks. We just don't support the purchase of nuclear submarines, and we hope that it won't materialize, because it doesn't have any benefit to us and it doesn't seem to contribute to the peaceful expressions and interests of our people.

Huddleston
I would just make a couple of points. I think there is a certain degree of questionable logic in continuing to attribute to a 1987 document the sentiments of people in 1989, when the intervening years have seen some very dramatic changes. The purpose of the 1987 White Paper was to describe the situation as the government then saw it. If you read the 1988 update by the Department of National Defence, you will see that there is a

199

significant difference in the description of the situation, and if you were to read the Department of National Defence update that is currently in preparation in 1989, you would see again a reflection of actual events. And of course people in uniform deal with actual events, not with optimistic dreams. The Department of National Defence update of 1989 will not speculate on the world of 1991, for example, any more than the White Paper of 1987 speculated on the world of 1989. I think that the most important issue for us now—aside from promoting these cooperative ventures that have been described in great number over the last day and a half—the most important thing to ensure that the current relaxation of East-West tension continues is that Mr. Gorbachev succeed at home. He seems to me to be the latest example of those famous prophets who have more success abroad than in their own country, and that is not difficult to understand, but it is a fact. There are tremendous challenges facing him. I think if he could get a couple of Loblaws stores in downtown Moscow to replace the current foodstore #49, he would go a long way towards assuring us that the type of world he has advocated is really going to transpire. We all hope that it does, and I certainly associate myself with those who do.

Roche

I think an indication of the evolution of thinking going on within the Canadian government today was revealed in the speech by Mary Collins yesterday. Moreover, I indicated some moments ago that the Consultative Group on Disarmament and Arms Control Affairs, which is a formal structural advisor to the government on these questions, at its most recent meeting some weeks ago said that, given significant adjustments to Soviet foreign policy and military doctrine over the past several years and the opportunities thus presented, there is a pressing need for a reevaluation of the Canadian approach to security at both the conceptual and policy levels. So there is a lot going on. Finally, to Dan Heap I want to say in answer to his question about who the enemy is: Dan, the enemy is the nuclear arms race and it is absolutely essential that all parties get onto common ground so that we can decelerate as rapidly as possible.

Oliver Jon Irlam (Youth for Peace, St. Albert, Alberta)

How does the military, representing themselves as good corporate citizens concerned about the impact of their activities on the environment and the people, explain their lack of concern over the results of low-level flight training in the Goose Bay, Labrador and Quebec area—those results being disruption of the way of life of the Innu, who have lived there for at least nine thousand years? And how can the low-level flight training of near-sonic jets not disturb the migratory patterns and habits of wildlife in the area?

Huddleston
I know that there are a lot of questioners, so I shall try to respond more briefly than I did to the previous one. The low-level flying areas in Goose Bay have no permanent inhabitants. The people who hunt and trap there seasonally are invited to advise the military where they will be, and then they are avoided by aircraft. The George River caribou herd currently numbers over half a million animals. It is at the highest level, by more than ten times, that it has been in recorded history.

Irlam
What does that have to do with their migratory habits?

Huddleston
The migratory habits are not affected by aircraft, because aircraft avoid the herd. The herd is tracked. The herd is followed. The herd is avoided.

Irlam
You did say earlier that you have done tests over the herds themselves.

Huddleston
We have done that to observe what effect aircraft have on caribou, and it doesn't alter what I have just said, that we in fact avoid the herd. We follow the herd and we avoid the herd. We attempted the tests over the herd because we are obliged to pursue the federal environmental assessment review process in presenting to government the proposals for the development of Goose Bay. That process is in train at the moment, and you cannot do that without doing a certain amount of experimentation. Otherwise, we are all dealing without facts. When we have flown jet aircraft across the caribou we have found that they were disturbed very little. What disturbs the caribou in comparison is to fly a very light helicopter over them—that stampedes them. I'm sorry, but that is a fact.

Clement Leibowitz (Edmonton, Alberta)
In view of the fact that the Soviet Union is the designated enemy, or potential enemy, in many official Canadian publications, and in view of the fact that in response to a number of peaceful and friendly gestures emanating from the Soviet Union, it was said as recently as yesterday here that we should wait to make sure that the changes in the Soviet Union are irreversible, what then would it take from the Soviet Union as a minimum

measure to let Canada stop designating the Soviet Union as the enemy against which Canada must protect itself? What would it take?

Huddleston
My understanding is that Canada does not currently designate the Soviet Union as the enemy. That is something that has developed as a result of the fact that Mr. Gorbachev has renounced those assertions of his predecessors, that the object of his system is to dominate systems such as ours. He has made it quite clear that he does not hold those views. Consequently his system cannot be seen as being inimical towards ours.

Leibowitz
Do you understand that now the military in Canada considers the Soviet Union as good a friend as the United States and will reorient its military perspective in consequence?

Huddleston
I'm not sure that I got all of that. I think we have to establish, as we are starting to do, a relationship with the Soviet military. We don't know the Soviet military; we have had no such associations with them. That is a process that has to begin. We have worked with the U.S. military for many, many years. We understand them. I think your objective and ours ultimately are the same.

Garnet Thomas (Project Ploughshares)
General Huddleston, I wish there was an elected representative of your department to answer these questions, but you are on the hot seat today, I guess. Now that we have begun practice missions to intercept cruise missiles with F-18s, and since I am personally convinced that there is no effective method of intercepting an attack by dozens or hundreds of these low-flying, dodging missiles except by lobbing nuclear weapons in their general direction, are nuclear weapons now to be stored at our forward fighter bases or at Cold Lake and elsewhere? I have a couple of other parts to the question. Will only American fighters carry these weapons, with our fighters flying along as a kind of Robin to their Batman? You stated that we would attack enemy bombers before they launched their missiles, and I am wondering how we would recognize a hostile approach as opposed to a normal conventional approach up to their usual fail-safe point until they had launched the missiles. If I am incorrect and there is an effective method of interception that does not involve nuclear weapons, would you lay it out for us in general terms? Finally, what was or is the interest that justifies ignoring the thinking of the majority of Canadians regarding the testing of cruise missiles?

Huddleston
If I may address the last part first, I will hide behind your initial observation that I am not an elected representative of government and therefore I am not really in the business of determining whether government will or will not respond to polls.

Briefly, all this stuff about using nuclear weapons to intercept cruise missiles is dreamland. There is no such tactic. There is no such effective tactic. There are no nuclear weapons stationed in Canada. There are not about to be nuclear weapons stationed in Canada. We obviously consider the Bear H and Blackjack[4] and other bombers armed with cruise missiles as a potential threat. We have a North Warning System whose primary purpose is raid warning and detection. We also have fighters with which we attempt to reflect the sovereignty of our airspace in peace time. We would employ the very small numbers of fighters that we have should a raid be detected, and we would attempt to intercept the bombers.

We know full well that nuclear-war fighting is not a viable option and that we don't have enough fighters to intercept all the bombers and cruise missiles in the world. I think I have answered at least eighty percent of your question.

John Plaice (University of Victoria, Victoria, British Columbia)
I came to this inquiry to learn more about the problems of peace and security and how they relate to the Arctic. What a surprise! The Minister of Defence of Norway in twenty-five minutes described in some detail the problems that must be addressed in the area of arms control, including some concrete proposals on the directions to be taken. On the other hand, Canada's ambassador for disarmament, also in twenty-five minutes, said nothing, just bland remarks about things done by nongovernmental organizations. Ambassador Roche, why don't you make any concrete proposals? Is it that Canada is just an American puppet?

Roche
Look, the Arctic, in the concepts of peace and security that this inquiry is trying to address, can only be seen in terms of East-West relations today. It has to be a part of the dynamic that is happening in the world. For a day and a half here I think that there has been a lot of information put out to give everybody a very strong sense of hope in what is happening.

There is an immense amount of cooperation by the eight Arctic nations that include the two superpowers, the Soviet Union and the United States. We are going through an unprecedented period of finding out how to cooperate. There are all manner of things going on now of an environmental and an economic nature to deal with the non-military aspects of cooperation. We are going to start exploring military aspects of cooperation. Mr.

Question Session

Gorbachev has made some important moves. So have we in responding to that. I think that, as the negotiations continue in Vienna particularly—not to mention in Geneva and at the United Nations in New York—the international community is coming onto some common ground, and the Arctic is going to be a beneficiary of this new thinking and new understanding. So I plead with this audience to send out a very strong message to all parties concerned, that the Arctic has to be factored into this global thinking for this new kind of world that we are now on the edge of.

Irene Gilmour (Qualicum Beach, British Columbia)
First I'd like to say that I used to live in the Northwest Territories. I lived there for thirteen years, and as did a previous speaker, I wish there was an elected official of the Canadian government to address this question to. I believe the Norwegian minister stated that sea-launched cruise missiles and attack submarines should be the major concern of an overall arms reduction strategy. General Huddleston, do you concur with this, and if so, why should Canada purchase attack submarines?

Huddleston
I think that Mr. Holst provided a very imaginative and intellectually challenging presentation covering just about every aspect of arms control in the East-West context imaginable, and he did make the two points you have just raised. My own attitude towards arms control is that one cannot treat it like a shopping list. All the pieces have to fit together. I propose, as one of my first actions on return to Ottawa, to get a copy of exactly what he said and read it and attempt to understand it better than I was able to in the very short time in which this intensive presentation was made.

I will deal with the two points specifically. Canada has made it quite clear that it would like to see sea-launched cruise missiles (SLCMs) included in the Strategic Arms Reduction Talks negotiations. We recognize that there are problems with that—verification problems because SLCMs can be conventional or nuclear—but that is the Canadian position. On submarines, there is a Canadian position that differs from that. It's a fascinating debate, but it isn't our position; and the government is not about to change its position at this moment. Let's continue that debate.

Kim Zapf (Edmonton, Alberta)
General Huddleston, as a Canadian who has also lived in the Northwest Territories and in southern Canada, I have, I think, a very simple question. I have heard from you this morning on issues of strategy and security, and I get the sense that the enemy is somehow out there in foreign countries. When I hear from Dennis Patterson I get the sense that the enemy is the foreign system in our own country. I would like to know, from

your perspective, clearly, who is the enemy, who is the greatest threat to our system? Whom are you protecting us from? I would be interested in a similar answer from Dennis Patterson.

Huddleston
I believe that there are many enemies. Anything that threatens our way of life, anything that threatens the society we have come to cherish is a threat. In recent times, environmental issues have been seen as such a threat. In times gone by we have tended to focus on, or think of, strictly military things when we use the word "threat"—at least in my world we have. Now we think of much wider threats. We think of medical threats: somebody mentioned AIDS this morning. Look at the threat that that poses to some African nations, for example. But in military terms we do tend to associate the word "threat" with military forces belonging to nations that have traditionally held views threatening to us. We see that changing. I think if one were to review the proceedings of the last couple of days, one would get a distinct sense that military threat is seen to be declining and that other threats seem to be increasing.

Patterson
The Northwest Territories being a territorial jurisdiction—without any real responsibility or many opportunities to participate in sophisticated debates on global military strategies—our people have simple values. They are close to the land. They love the environment. They are dedicated to good relations among circumpolar neighbours. And they have said very strongly in our legislature that they are against what they see as Canada's contribution to the acceleration of the global arms race. We're against cruise-missile testing. We're against low-level flying exercises. We're against nuclear submarines in our waters. So I would say that our enemy is the global arms race and whoever we see participating in that destructive course.

Bob Tennant (Calgary, Alberta)
General Huddleston, it is quite evident that the United States is a significant problem itself in the developing menace of northern militarization, and it seems that you have ignored the role of the United States in militarizing the Canadian Arctic. Why do you ignore this role to the extent that you have?

Huddleston
I wasn't conscious of the fact that I had ignored it. I spoke of the North American Aerospace Defence Command (NORAD) at some length, which as you know is an arrangement binding Canada and the United

States, so that, implicit in anything about NORAD, one includes the United States. We talked about cruise-missile testing. We all know whose cruise missiles we are speaking of. I'm really not conscious of having avoided reference to the United States. I'm sorry.

Art Smith (Peace Network, Wetaskiwin, Alberta)
Why are we involved in arms buildup, nuclear submarines and sophisticated systems at all? I came here to hear about peace. Many of the speakers so far, especially the armed services personnel, seem to have two wrong attitudes on which they are basing their assumptions: first of all, that we have an enemy, and second, that the enemy is going to start a nuclear war. Why won't we let the superpowers play their $100,000 million kindergarten war games, and let us not get caught up by spending our $116,000 million on nuclear submarines. We know there is not going to be a nuclear war, so why do we pretend that there is going to be so that we can build up our arms in Canada?

Huddleston
I do not believe that people in uniform have the attitudes you described. It's rather a well-worn phrase to say that we regard ourselves as the largest peace force in the country, but in fact we do, and I'm not simply referring to peacekeeping activities throughout the world, which of course we engage in as much as any other country. We do not have those attitudes.

I do agree with you that nobody is going to start a nuclear war. I think the reason that they are not going to start any kind of war, *any* kind of war, is that, at a time when political and ideological differences have set nations against nations, we have a form of security that deters the use of war as a means of achieving ends. Now we are on the threshold, I hope, and I share that hope with you, of a time when the need for that will diminish.

Roche
Thank you, Adrienne, for allowing me to comment. It's not that I am feeling neglected up here, but I think that maybe General Huddleston, because he is wearing a uniform, seems to be an object of either affection or something else.

In the final analysis, the Canadian government is all one entity in its political structure, which leads me to make a parenthetical note—Adrienne, I hope you will allow me.

I see two elected politicians—and I apologize if I haven't seen any others—two federal politicians, Audrey McLaughlin and Dan Heap. Audrey is from Yukon and Dan is from Toronto. I would like to know where the Alberta members of Parliament are. It seems to me that a lot of these members of Parliament, who are in fact not sitting in Parliament at the moment, might have got a lot out of coming to these two days, as I myself have.

Now, about the military; there is an underlying current here that makes me just a bit uneasy, in some of the questions—I put this to you respectfully—that Canada's military is some sort of bogey man that is the cause of all the problems in the world. It is quite the reverse. From my perspective—and I do a lot of work at the United Nations—of 144 reporting nations, Canada is ninety-fourth in terms of what we spend on the military related to our gross national product. We are so small you can't even find us on the military map most of the time. As a matter of fact, you could put Canada's entire Armed Forces in McMahon Stadium in Calgary.

And, speaking of Canada's Armed Forces, what about Canada's role in peacekeeping? We have had eighty thousand participants, Canadian Armed Forces participants, in fourteen peacekeeping missions around the world. This is so important that the United Nations won the Nobel Peace Prize for peacekeeping work. Granted there are a lot of other countries involved too, but Canada is certainly at the forefront. How are we going to do peacekeeping if we don't have a military?

We do have a military beyond peacekeeping. We have a military because this government under its political, democratic, elected structure has a responsibility to the Canadian people to have a defence system. The Canadian military have performed that role with distinction over the years and I think they ought to get some credit for it.

Notes

1. Negotiations on Conventional Forces in Europe (CFE), in the context of the Conference on Security and Cooperation in Europe (CSCE), had recently begun in Vienna at the time of the inquiry. (Editor)

2. Several families were relocated from Inukjuak (Port Harrison) and Pond Inlet in the early 1950s to these two high Arctic localities, ostensibly to take advantage of better employment and hunting opportunities and allegedly to strengthen Canadian sovereignty there. A number of the people have since been returned to their home towns, at their request, by the federal government. (Editor)

3. "Challenge and commitment: A Defence Policy for Canada" (Ottawa, 1987).

4. Lt.-Col. Murray R. MacDonald describes the Soviet Bear H bomber as an updated version of the Bear, equipped with six AJ-15 cruise missiles with a 3,000-km range: there are over 50 Bear Hs in the Soviet fleet. MacDonald describes the Blackjack as a newer supersonic long-range bomber. See "The air and space dimension of the threat" in "Peace and security in the Arctic," Report of the Consultative Group on Disarmament and Arms Control Affairs, October 1987 (Canadian Centre for Arms Control and Disarmament: Ottawa, 1987). (Editor)

Soviet Approaches to Security and Peaceful Cooperation in the Arctic

Ambassador Alexei A. Rodionov

CHAPTER 19
Soviet Approaches to Security and Peaceful Cooperation in the Arctic: An Overview

AMBASSADOR ALEXEI A. RODIONOV

First of all, I would like to thank the organizers of this inquiry for their kind invitation to Edmonton. The presence of the Soviet delegation at this forum testifies to the importance attached by the Soviet Union to the problems of the Arctic and the North.

The Soviet Union believes that the Arctic is of enormous global importance. It is in the common interest of humanity to transform it into a region of mutually beneficial cooperation between all interested parties and not into a playground for hostilities and military sabre-rattling.

We believe that the positive developments presently taking place in the world create favourable preconditions for drafting a viable formula for the stable and secure development of the Arctic. Contrary to certain allegations, we are not trying to take the problems of the North out of the general context of international security. The Soviet Union supports a comprehensive approach to the issue of ensuring global peace. Our well-known

Alexei A. Rodionov is the Soviet Union's ambassador to Canada. He is an economist, a graduate of the Moscow Financial Economics Institute and the Diplomatic Academy of the U.S.S.R. Ministry of Foreign Affairs. He has been in the diplomatic service for twenty-seven years and an ambassador for approximately twenty-one of them. He was Soviet minister-counsellor to India and ambassador to Burma, Pakistan and Turkey before his posting to Ottawa. He was also Minister of Foreign Affairs of the Russian Soviet Federated Socialist Republic and a member of the collegium of the U.S.S.R. Ministry of Foreign Affairs. He has participated in a number of international conferences and was a member of the Soviet delegation to the General Assembly of the United Nations.

concept of an all-embracing system of international security is a vivid example of this approach.

However, we are convinced that at present there exist objective prerequisites for creating a model of fruitful and mutually beneficial relations in the North, which could become an example for other regions in the world.

As is known, the Soviet Union's considerations regarding such a model of relations were put forward by Mikhail Gorbachev in his Murmansk statement in October 1987, which called for transformation of the Arctic into a zone of peace and cooperation. We regard the creation of such a zone as a complex process that includes parallel steps along the following directions: lowering the level of military confrontation in the region; building confidence between Subarctic and northern states; expanding cooperation in the field of Arctic natural resource development; conducting scientific research, including research aimed at keeping an ecological balance in the region; and ensuring the social and economic rights of the native population.

We regard the creation of a zone of peace in the Arctic as a stage-by-stage, progressive process towards consistent implementation of measures in all these fields, on both a bilateral and regional basis. We call upon all interested parties to develop an intensive and multidimensional dialogue through diplomatic and political channels between governments, representatives of political parties, scientists, businessmen and social organizations.

Life itself urges the Arctic nations to pool their efforts. Let us take ecology as an example. The problem of protecting the environment in the Arctic is acute. One has to honestly admit that all the northern countries have made their contribution to the pollution of the Arctic waters, and they are not the only ones. We of course have contributed our share of pollution too. But today the issue is not who did more or who did less damage to the ecology of the Arctic. The issue is how, through joint efforts, to prevent this process in the future and how to alleviate most effectively the damage already done.

What are the main ways to eliminate the consequences of environmental pollution in the Arctic and to prevent it in the future? Here are some of them:

• Consistent adherence by the states to their international commitments in the sphere of ecology
• A movement towards ecologically clean resource development and energy-saving methods in the economic development of the Arctic
• Joint development of quality criteria and standards for the environment in the North

- Joint control by the Arctic countries over the state of the Arctic environment
- Exchange of advanced technology to utilize the natural resources of the Arctic and of scientific research in the field of Arctic ecology
- Dissemination of ecological knowledge and culture
- Regular exchange of information among the Arctic states about their activities in the field of ecology and their application and improvement of national legislation in this field
- Adoption of joint practical measures for the protection of the Arctic environment

In his speech in Murmansk, Gorbachev proposed to develop jointly a common, complex plan to protect the environment in the North and to set up a system of control over the state of the environment and radiation security in the region. Our experience using collective measures to protect the marine environment of the Baltic Sea could be extended to all the oceanic and sea areas of the North.

We know that other countries have interesting ideas to this effect also. For instance, a conference on the Arctic environment as proposed by Finland would, I think, be very useful. I would like to express my special gratitude to Ambassador Esko Rajakoski for his elaboration of this proposal. We find also very promising the exchange of opinions that took place at the conference of the Subarctic countries in Leningrad in December 1988 on coordination of scientific research and at the meeting of parliamentarians of these countries in Moscow in January 1989 on the problems of northern ecology. For this reason, we believe the practice of holding such meetings should be continued.

The Soviet Union pays a great deal of attention to cooperation in the development of Arctic resources. It is necessary, in our view, to establish jointly a mechanism for such cooperation in the interests of rational development of northern regions. The importance of close cooperation, the exchange of information and the coordination of scientists of various countries in the field of northern research is hard to overestimate. Mr. Kazmin will speak in detail about the Soviet approach to possible ways for interaction in these fields.

I would emphasize once again that on the whole we see good prospects for developing cooperation between Arctic and northern states on a wide range of scientific, ecological, economic and humanitarian problems of the Arctic. But it is highly unlikely that we will reach, to the full extent, the potential of peaceful cooperation if we have to, figuratively speaking, elbow our way through the armada of military vessels, many with nuclear armaments, that is now concentrated in this region. The militarization of the northern seas is dangerous, in our view. It destabilizes the overall military strategic balance in this area and poisons the political climate.

In Murmansk we proposed that all interested states take part in negotiations on the limitation and reduction of military activity in the North in both Eastern and Western hemispheres. This initiative was further developed and given a concrete character. Unfortunately, our initiatives have not received a positive response from the North Atlantic Treaty Organization (NATO). We hope that this is not its final word. We expect at least counterproposals from NATO.

The necessity to lower the level of military confrontation and to enhance military and political stability in this region stems from the logic of ensuring the security of the Arctic as a whole and the security of each of the northern states. We are convinced that this necessity is both a prerequisite to and an integral part of transforming the Arctic into a zone of peace and fruitful interaction. Mr. Granovsky, our expert in this field who is participating at this conference, will speak on this issue more specifically.

The Soviet Union by no means sees the implementation of the military aspects of the Murmansk initiatives as a compulsory precondition for any movement in the civil areas. We are prepared to deal with any issues that could be solved today. In this regard, I was encouraged by the interesting remarks on confidence-building measures at sea made by the Minister of Defence of Norway, Mr. Holst. We will take these proposals and ideas into consideration.

We consider the Murmansk program a starting point for the long and complex process of establishing a reasonable *modus vivendi* in the North. This process will include new initiatives that take into account the positions of other countries, their concerns and suggestions. But to take their views into consideration, we need to hear from them first. That is why a dialogue on a whole range of issues, including the military issues, is so necessary.

Our call for intensifying multilateral cooperation in the Arctic does not mean at all that the Soviet Union is losing interest in further interaction with its northern neighbours on a bilateral basis. We will continue to pay the most serious attention to promoting this interaction with those countries, naturally including Canada.

As ambassador of the Soviet Union to Canada, I would like to note with satisfaction that Soviet-Canadian cooperation in the Arctic is gaining momentum on the basis of the program of scientific exchanges signed in 1984 and broadened in 1987. We believe there is a clear and objective need for our two countries to expand this cooperation further, taking into account the similarities of tasks that the Soviet Union and Canada face in the field of Arctic and northern development. We are interested, for instance, in the participation of Canadian companies in joint ventures for developing oil and gas deposits in the shelf of our northern seas. The decision to conclude an intergovernmental agreement on Arctic cooperation between Canada and the Soviet Union opens promising new perspectives in this regard.

CHAPTER 20
Soviet Approaches to Security in the Arctic

ANDREY E. GRANOVSKY

I can say frankly that I am a bit nervous about the time limit on my address, because so much has been said during this one and a half days about the Soviet approach to military problems of the Arctic that it is very difficult just to react to these thoughts or to reflect on them in this short period of time. I have just finished rewriting the information I will give you.

Before turning to a description of our position on military questions, I would like to clear up one small point, or rather factual mistake, that I heard yesterday. The "comprehensive approach" to international security was not at all, as was said here, a longstanding Marxist position that Gorbachev's regime abandoned. Unfortunately, our Marxist approach to security was identical to the capitalist approach to security, in the sense that the approach of both was that security must be ensured by military means—by strong armies, strong navies, etcetera. Actually, it took us decades to become mature enough to realize the obvious fact, which today seems to be lying on the surface, that real security cannot be achieved by military means.

Andrey E. Granovsky is counsellor, Department of Arms Limitation and Disarmament, Ministry of Foreign Affairs of the U.S.S.R., in Moscow. He is a graduate of the Moscow Institute of International Relations and holds a Ph.D. in political science. He began his career with the diplomatic service in 1975, serving for most of the next ten years in India, and subsequently moved to the Press Department, before accepting his present position. He has participated in several disarmament negotiations and is the author of several articles on arms control and security matters.

And it is with *perestroika*, and only with *perestroika*, with a fresh look at ourselves, what we are, how we live, and the surrounding world, that we have advanced a comprehensive approach towards international security. This approach holds that security lies in the political sphere, not in the military one. What does it mean, in two words?

First, national security is an integral part of the concept of equal security for all and not otherwise. No country can be more secure than others. We should be interested in the security of the other side as much as in our own, because our partner's insecurity creates mistrust, suspicion and, in effect, insecurity for everybody. This is why we stand for elimination of all imbalances and asymmetries in armed forces, including those spheres in which we have superiority.

Second, real security can be achieved only through mutual trust and respect and the interest of one side in the other. Thus, the way to security is reached not only through disarmament efforts, but also through cooperation in solving common problems—ecological, economic, social, humanitarian, etcetera. This is why we call our approach a comprehensive approach to security.

Our idea of turning the Arctic into a zone of peace and fruitful cooperation is a reflection of the general philosophical concept of security in our regional policies. We consider that substantial lowering of military confrontation—including nuclear confrontation—in the Arctic and in the areas adjacent to it is an integral part of turning the Arctic into such a zone. However, despite any suggestion that there is a linkage between military and non-military aspects of our Murmansk program, the fact is that we have never considered limitation of weapons in the region as a precondition for development of cooperation in other spheres of action.

As Ambassador Rodionov said, we are in favour of solving those problems that can be solved today, as a general amelioration of the political climate in the region will create more favourable conditions for the solution of more complicated problems. However, we are sure that the existing concentration of arms in the region gives weight to the spirit of confrontation and mistrust and hampers the development of cooperation in other spheres.

[Editor's note: Passages in italics were in the written speech but were not delivered orally because of time constraints.]

Naturally, of greater concern to us is the area adjacent to the Arctic from the European side. In this connection, Finland's idea of creating a nuclear-free zone in Europe is important to us. Should the northern European countries decide to create a nuclear-free zone, the Soviet Union would be prepared to become its guarantor. We also agree to discuss with the northern European countries measures—quite substantial measures at that—with regard to the Soviet territory adjacent to a nuclear-free zone in northern Europe.

In our view, it is important to start practical work for creating a northern European nuclear-free zone, to begin a diplomatic process and an interstate dialogue. However, the impulse must naturally come from the northern European countries themselves as participants in a future zone. For our part, we are prepared for the broadest contacts in relation to the zone in various fields, including the military field.

I now turn to our position on military problems. The Soviet Union has advanced a whole complex of proposals to diminish the level of military confrontation in the North. We propose, for example, to all states concerned, that talks should begin on limiting and reducing the scope of military activity in the North as a whole, in both the Eastern and Western hemispheres.

A question was raised in this inquiry about why the proposals by Gorbachev do not cover the Barents Sea. Actually, it is not correct to say that they do not. We have clearly said on many occasions that we are ready to include the Barents Sea in the area of confidence-building measures. As for limitations of naval activity, we believe that limitations should be mutual. We have an area where there are fleets of North Atlantic Treaty Organization (NATO) countries and also of the Soviet Union—this includes the Norwegian Sea, the Greenland Sea, the North Sea, etcetera; this area presents a good possibility for reduction. There is a Soviet fleet in the Barents Sea, and there are other fleets in the Western Hemisphere, so one should take into consideration that the reductions must be balanced.

We give two options. One is to solve the problem of the Arctic as such, both Eastern and Western hemispheres. That would be preferable to us. The second, as a stage, as a first step towards this goal, is to speak about a concrete reduction of the tensest military confrontation that exists now.

We propose that all states concerned begin talks on limiting and reducing the scope of military activity in the North as a whole in both the Eastern and the Western hemispheres. As a step towards those talks, we propose consultations between the Warsaw Treaty Organization and NATO on reducing military activity and limiting the scope of action of naval and air forces in the Baltic Sea, the North Sea, the Norwegian Sea and the Greenland Sea, and also on extending confidence-building measures to those areas.

In order to prepare for the consultations between NATO and the Warsaw Treaty Organization that we propose, we have suggested that military experts of the two military-political alliances come together and discuss the problems.

We are also in favour of an agreement prohibiting naval activity in mutually agreed areas of international straits and generally along the lanes of intensive navigation; we have made a proposal to reach such an agreement.

We also propose the following:

1. To limit the number of major naval and air force manoeuvres in the foregoing areas to one manoeuvre every two years. We believe that that is quite sufficient, and that the scale and intensity of manoeuvres today on both sides are not justified by such considerations as keeping fleets in order and especially do not go with the existing climate in Europe and in the world today.

2. To create in the North and West Atlantic, for both the Soviet Union and the United States, agreed areas where anti-submarine forces and the forces of the military-political alliances will be prohibited.

3. To renounce on a mutual basis the carrying out of naval manoeuvres in the major ocean and sea commercial waterways in the North Atlantic and in areas of intensive seasonal fishing.

4. Not to allow naval force buildups in or near international straits and also to define the ultimate parameters of such groupings in terms of the numbers, types and other characteristics of ships allowed.

5. To include in the limitation areas the straits used for transits between the Baltic and North Seas and the English Channel, and the area of Iceland/the Faroes/Scandinavia.

These are our proposals. They naturally reflect our own understanding of how it is necessary and possible to advance toward greater military-political stability in the North. At the same time, we do not claim that we possess the ultimate truth. We are ready to discuss, in a very constructive manner, any other proposals. We expect that our partners will not simply reject our initiative in the military sphere but will be ready to search jointly for ways to increase stability in the region. We expect their realistic counterproposals, because only on this basis is it possible to find a balance of interests between the two alliances and other countries and to work out mutually—I underline mutually—acceptable steps towards greater stability in the region. There is no other way to this goal but dialogue.

The most typical argument advanced by critics of the Soviet proposals on reducing military confrontation in the North—and it was repeated here several times—is that the region should not be considered in separation from the overall global context of military confrontation, including nuclear confrontation, as neither our nuclear forces nor our navies are linked to a particular region. Actually, we do not see any contradiction between a regional approach and a global approach. Each individual problem should be solved where it is easiest to solve, whether on the global level or on a regional or subregional level. But to define it we need dialogue, not just to condemn each other publicly.

In connection with this argument, I would like to outline our position more precisely. First, as to the strategic nuclear forces, nuclear-powered strategic ballistic-missile submarines (SSBNs), stationed in the North, they are indeed global in character and are discussed at the Soviet–United States nuclear and space talks in Geneva. We expect other states that possess strategic nuclear forces to join the next phase of the process of their reduction. However, nonstrategic navies are not yet included in the framework of any disarmament talks, although it is surely incorrect not to consider this important component of the armed forces, which is powerful and universal in use within the overall military balance. The Soviet Union has repeatedly advanced various options for global negotiations on the navies, but as yet those proposals have not elicited any response from our partners. All these proposals still stand. They concern the limitation of sea-based tactical nuclear weapons, the prohibition of sea-launched cruise missiles, etcetera.

Meanwhile, given the lack of readiness on the part of the West to engage in global negotiations on the navies, the Soviet Union submits that we should start to resolve these major global problems at regional levels. This approach seems to us a practical one. It is astonishing that, after our proposals for global negotiations have been rejected, when we have submitted proposals to begin from regional levels, the reaction is that these questions are not regional, but global. All right, let us try to solve them on a global level. We stand for global reduction of naval armaments down to the level of reasonable sufficiency for defence; that is, to a minimum level at which navies would not be capable of delivering a surprise strike or of carrying out wide-scale offensive operations.

Yet our partners clearly show that they are not prepared to go along either with balanced reductions or even with limiting the navies. One may question the logic of this position; I mean the contrast between proposing the elimination of imbalances in the armies in Europe, in which the superiority is ours on one side, and refusing to engage in negotiations on reducing or even limiting the navies, in which NATO is evidently superior on the other. But the fact remains that it takes both sides to reach an agreement, and there is no consent yet.

Bearing this in mind, we propose taking measures in northern Europe that would not interfere with the existing structure of navies or with their development and modernization programs, but which at the same time would lend more stability to the region, which is not at all safe. These include measures to limit naval activities—not a reduction of fleets, but a reduction of naval activities—in the region; the scale of those activities not being consonant with the current international climate. We also propose to embark on confidence-building measures.

We are prepared to go ahead step by step and to start with the simplest confidence-building measures, such as, for example, regional agreements on prevention of incidents in the open sea. While the question of whether the regional approach to naval reduction is justified or not is debatable, there can be, in our opinion, no arguing over the usefulness of confidence-building measures. They constitute a normal element of civilized interstate relations of mutual respect, which we all are now trying to establish. These measures minimize suspiciousness and the danger of an incorrect perception of the other side's actions and subsequently of an incorrect reaction to these actions. That is very important.

We are convinced that greater openness in naval activity will promote better predictability of the situation in the North. We propose in this regard that nuclear states, that is, the Soviet Union, the United States and other nuclear states, declare whether their war ships calling at foreign ports have nuclear weapons on board. We also support the idea of the joint working out by the interested countries of technical means of verifying the absence of nuclear weapons on board. That is important in the context of future nuclear-free zones, in particular in the north of Europe.

Furthermore, much has been said here by representatives of NATO countries about the offensive character of the Soviet navy. I shall not follow suit, but the fact is that both sides consider their own naval groups in the North to be defensive in nature, while the other side's groups are perceived as offensive. With that, both have serious doubts about the truthfulness of the other side's statements, and this produces mistrust. So why do we not stop publicly condemning each other and charge our experts to try to reach a common or better understanding of the existing balance of naval forces in the northern regions, to get a better idea of each other's objectives in the region and of each other's concerns?

For this it would probably be necessary to compare data on naval potentials of both sides in northern Europe and in the North in general, to discuss on a factual basis the naval structures of the Soviet Union and of the United States and the other NATO countries, the designation of those forces, the correlation of different naval functional components, and the character of war games and manoeuvres. On the basis of such an analysis, we could jointly reflect on the changes that naval groups of both NATO and the Soviet Union in the North have to undergo in order to become strictly defensive in nature. We suggest that these topics be discussed either within the framework of the consultations between the Warsaw Treaty powers and NATO or between their military experts or anywhere. We are quite open here. In our view, the mere existence of such a dialogue would promote trust between the two sides.

Andrey E. Granovsky

The Soviet Union agrees with the notion that peace and security in Europe and in the world are inseparable, and we look at strengthening stability and security in the Arctic and in the North in general in the context of consolidating international security. But one more conclusion can be drawn from this postulate of inseparability, that is, that ongoing positive processes in Europe and in the world should either embrace or positively influence the situation in the North. We are sure that at present the most favourable conditions exist for bringing such changes about, and we should make use of these conditions.

CHAPTER 21
Soviet Approaches to Peaceful Cooperation in the Arctic

YURI V. KAZMIN

At the outset I would like to thank the organizers of this inquiry for the opportunity to come here to Alberta, to Edmonton. I had a chance to talk to some commercial people and government people from Alberta in October last year, and I got an impression that all the best people are in Canada, and that all the best in Canada are in Alberta in Edmonton. So I had a good opportunity to talk to these people, and I appreciate the efforts of the people of Alberta and Edmonton who have organized this very important meeting to discuss the very important problem of the Arctic. This will give other people the opportunity to understand that not all are bad guys in the Arctic, and that there are many problems that should be taken care of, not only by the people who live there but by the whole community on both a national and international basis.

My government pays great attention now to Arctic affairs. The evidence of that is the establishment of the special government body, the State Commission on Arctic Affairs of the Council of Ministers of the Soviet Union, which is headed by the alternate member of the Politburo, the First Deputy Prime Minister of the government of the Soviet Union, Mr. Masljukov.

I would not want you to leave this inquiry with the impression that the Soviet Union engages in devil tactics with respect to cooperation, as you might gather from some of yesterday's remarks. The government of the Soviet Union has advocated a long-term policy that international cooperation is a vital part of peace and security in the Arctic. For instance, in the

Yuri Kazmin is the Deputy Director for the State Commission on Arctic Affairs of the Council of Ministers of the U.S.S.R.

early 1970s some agreements were signed on a bilateral basis, which included cooperation on some scientific research and environmental protection in the Arctic. As an example, I will cite the Soviet–United States agreement of 1972 on world ocean research, which included some scientific research dealing with the Arctic. Also it is the case that the agreement between the Soviet Union and the United States in 1972 on environmental protection has a special reference to Arctic and Subarctic ecological systems.

I will also cite as an example the agreement between the Soviet Union and Norway on the management of fisheries in the Barents Sea. It was a good example of cooperation on a bilateral basis for the utilization of natural resources in the area, taking into account particularly that the Soviet Union and Norway at that time had some political problems, which are not yet resolved, with respect to the delimitation of their economic zones in that area.

Of course, the Murmansk speech by Mr. Gorbachev, which was referred to by many speakers yesterday and today and which was elaborated upon by our ambassador this morning, put Arctic problems on the political agenda and provides a good opportunity for finding ways and means to improve cooperation on a bilateral and multilateral basis as a further contribution to developing peace and security in the region. Since that speech some progress has been achieved with respect to the development of that cooperation.

Our country has moved in several fields. For instance, in 1988 we signed an agreement between the Soviet Union and Norway on scientific and technical cooperation in the Arctic, which includes some fields of scientific research, dealing with geophysics, geology, oceanography and Arctic biology. Also in 1988 a new agreement on cooperation in the field of environmental protection was signed with Norway. You have already heard much about the broadening of cooperation on science and technology between the Soviet Union and Canada, so I will just express my thanks to the previous speakers who have made my task easier in that respect; but I do want to point out that the new agreement, which is drafted, includes cooperation in the field of socioeconomic problems of the native peoples of the North, including cultural and academic exchanges and contacts between the native peoples.

With respect to bilateral arrangements for cooperation between the United States and the Soviet Union, some steps have been taken to include Arctic items in renewed agreements on ocean research; and Soviet and U.S. cooperation on Arctic matters received favourable mention in the joint communiqué issued during summit meetings in December 1987 and June 1988.

Looking more closely at the fields of cooperation proposed in Mr. Gorbachev's speech, I would like to mention that cooperation in resource development of the Arctic and the North may include, for instance, cooperation in the development of petroleum resources, onshore and offshore. We have some talks going on with the Arctic and Nordic countries for development of offshore natural gas deposits and for development of onshore oil and gas deposits in the northern part of the European districts of the Soviet Union. Talks about cooperation with respect to the management of the resources of the Kola Peninsula—I mean hard minerals, heavy minerals and phosphates—are also under way. That cooperation with respect to the Kola Peninsula has a good influence on the problem of environmental protection, because it provides for the modernization of already-functioning refinery plants in order to protect against air pollution. The next step is to construct new production facilities to utilize the hard minerals without creating wastes and to develop new advanced technology for protection against air pollution.

Our ambassador has already told you much about our thinking on cooperation in environmental protection. As well, of course, the proposal we heard yesterday from the Finnish ambassador, in my opinion, should be considered favourably by our government. I think that this proposal shows what a country, whether it is big or small, may accomplish if it wants to do something and not just to talk about the problems.

Besides the signing of agreements on the subject, there has been some actual cooperation on environmental protection. I have mentioned the visits of scientists and specialists on environmental protection from Nordic countries to the Kola Peninsula to understand better what that task entails. I may also cite as an example of cooperation on environmental protection the joint Soviet and U.S. effort in October 1988 to save three, though unfortunately it came down to two, whales trapped in the ice off the shore of Alaska.

With respect to scientific cooperation, I would like to underline the importance of the Leningrad conference of the Arctic countries on coordination of research in the Arctic, which was held in December 1988. Some speakers have already mentioned the conference, but its importance is that during the meeting more than five hundred scientists from all Arctic and Nordic countries and some other countries discussed not only the problems of science in the Arctic but—and they put these in their final document—the areas in which cooperation should be carried out on a multilateral basis. If an international Arctic science committee is established soon, the final document of the conference would be a good guideline for that committee to follow in order to promote international cooperation in scientific research. That final document indicates the fields and the ways and means of dealing with the problems of scientific research. These fields include: the upper atmosphere and near space; Arctic ecosystems; interaction between the ocean and atmosphere on Arctic climate changes; geology; geography and glaciology; and environmental conservation. There is also a long list of recommendations on items in the field of socioeconomic, educational and cultural problems of the indigenous people of the North and on the problems of polar medicine.

In his speech, Mr. Gorbachev mentioned that cooperation should take care of the interests of the native indigenous people of the region. There are indications of progress in that respect. Examples are the exchanges of northern people between Canada and the Soviet Union, which you have heard about at this inquiry; the visit of people from the Inuit Circumpolar Conference (ICC) to the Soviet Union last year; and the readiness of our government to send a delegation to the forthcoming ICC general conference. This is a very important subject. We have to cooperate.

At the meeting of parliamentarians on the ecology of the North, a Canadian representative said that the native people in Canada have a big problem, because whenever there is a struggle between the economy and the interests of the native people, the economy wins. The same problem exists in our country, where there are twenty-six northern ethnic groups, fourteen of which live in the Arctic.

As a first step, the State Committee on the Arctic decided to analyse the problem. We have now decided to adopt the policy of promoting opportunities for both the development of traditional lifestyles of ethnic minorities and reasonable but not damaging economic development. We have, for instance, decided to carry out an inventory of local natural ecosystems, such as reindeer grazing, to identify the areas requiring reclamation and protection. We consider that we have to ensure real preferential rights for the indigenous Arctic populations to manage use of the land and natural resources in the area of their homeland.

I would like to stress that these policies, which we are adopting now with respect to international cooperation in non-military matters, actually represent the process that is going on in our country. It is a process that has been mentioned already by my colleagues. We call it *perestroika,* and with respect to foreign policy, it is known as "new thinking" or *novoje myshleneje.*

I would like to end my statement here by saying that in his address to the Soviet people on New Year's Eve, Mr. Gorbachev indicated that our struggle for our social aims and for the development of socialism in our country is connected with the need to resolve universal problems and that, in this sense, our *perestroika* is of universal significance. This places a vast historical responsibility upon us.

Yesterday the media asked one of my colleagues why we had come here. My reply is this: just to carry out our responsibility with respect to universal problems, we came here. This is the answer.

CHAPTER 22
Maritime Strategy and Nuclear-free Zones

CAPTAIN JAMES T. BUSH

I am a member of the Center for Defense Information. Please do not make the mistake of thinking that I represent the U.S. government, particularly the executive branch of the government, that is, the navy or the Department of Defense. Many of the things that I will be saying this afternoon will be in opposition to the beliefs of the Department of Defense of the United States.

I would like to say, however, that there are three branches of the U.S. government, and the opinions of the Center for Defense Information parallel the opinions of many members of our Congress—I would say the majority of our members of Congress. That is important.

For example, the Center was very influential in the passing of test ban legislation by our House of Representatives. This would have limited— severely limited—U.S. testing capability to one kiloton. The legislation just about passed the Senate. Had it done so, the Congress of the United States would have essentially negotiated a treaty with the Soviet Union. As well, Congress has stopped the U.S. government from testing anti-satellite weapons, and it is going to be very much involved in the question of modernization of nuclear weapons this year. Our Congress is very powerful.

To my mind, therefore, it is not fair to say that I do not represent the U.S. government. Because I do represent, I believe, a very large portion of one branch of that government, the Congress.

I did not realize before I came that I was the only American who would be speaking. Because of that, I have changed my speech slightly. I feel I should answer some of the questions that have come up here in the last day and a half. My speech is advertised as being a discussion of the mari-

Maritime Strategy and Nuclear-free Zones

Captain James T. Bush

time strategy of the United States, but I am going to accompany that with a discussion of the U.S. position on such issues as nuclear-free zones.

I will start with the U.S. maritime strategy. I am very happy that Mr. Holst outlined to you this morning the roles and missions of the navy, because what I would like to do this afternoon is to tell you how the U.S. navy has converted those roles and missions into a basic strategy—what they intend to do with the navy in a war.

The maritime strategy is an overall strategy that deals with peacetime and wartime. The strategy has three parts. The first part is known as "peacetime presence." During times of peace, the U.S. navy, with its roughly 550 ships, makes its presence felt throughout the world in an effort to threaten the world with this power and by that threat to deter war. Part of this peacetime presence is to go into those areas of the oceans that are considered to be international waters by the United States and considered to be home waters by the littoral states and very strongly make our position that these are international waters. Canada, of course, has been one of the countries that we have done that with the most. We also made a trip into the Black Sea and passed intentionally through Soviet territorial waters on that trip. This is part of the maritime strategy—to keep the threat of the navy ever present in the eyes of the world.

The second phase of the maritime strategy is known as "crisis response." If the United States sees a crisis anywhere in the world—that is the way it is stated—the United States polices the whole world and, if there is a crisis anywhere, the United States responds. The key to the crisis response is what is known in the buzz words of the navy today as "escalation control." At every level of armaments, the United States is superior to such an extent that our enemies will realize that if they esca-

Captain James T. Bush, United States Navy (retired), is an Associate Director of the Center for Defense Information in Washington, D.C. He entered the U.S. navy after graduating from the University of Michigan. He was selected for submarine duty in 1956 and for duty in nuclear submarines in 1960. He served first on USS Bluegill *and subsequently on the* Robert E. Lee, Triton *and* John C. Calhoun *before assuming command of USS* Simon Bolivar *in 1967. He later served as a staff officer with the Commander in Chief, U.S. Naval Forces Europe, in London, and as Operations and Readiness Officer for the Commander, U.S. Second Fleet, in Norfolk, Virginia. He served on the staffs of congressmen and senators and worked as a stockbroker before taking up his present appointment. Captain Bush holds a master's degree in international relations from the University of Southern California. While an active officer he received several letters of commendation in addition to the Navy Commendation Medal and the Navy Achievement Medal.*

late from a conventional war to a tactical nuclear war, they are going to be facing a superior enemy; if they go from a tactical nuclear war to an intermediate nuclear war, they will be facing a superior enemy; if they go to a strategic nuclear war, they will be facing a superior enemy. That is the essence of escalation control, and that is the essence of our maritime strategy in the crisis-response phase.

The third phase of the maritime strategy is the "war-fighting phase." Should crisis response lead to a war, then we have tactics for the war. At this phase we are talking about the Soviet Union, because anybody else would have been put down by our deterrent forces or by our crisis response or by our peacetime presence—nobody else is going to initiate a war with the United States.

The tactics for war are divided into three parts. The first part is deterrence. It is difficult to separate this from escalation control, but the idea is that we will, using our strong forces, deter an extension of the war. So we move our forces forward. During this phase or even in the crisis response phase—one of the problems with this maritime strategy is the vagueness of the line between crisis response and war fighting—as the U.S. forces move forward, we may start sinking Soviet missile submarines, despite the fact that the war has not escalated to a nuclear war.

The next phase of the war fighting is called "seize the initiative." In this phase, we have our forces all forward; we attack the Kola Peninsula; we are sinking Soviet missile submarines as much as we can; we are carrying the fight to the enemy. Very aggressive.

If the Soviets have not as yet been defeated, the final phase begins. We bring in everything we can, and we conclude the war on terms favourable to the United States. This is not a "war-winning strategy," you understand; it is "to conclude the war on terms favourable to the United States."

I would like to focus on the problems in this strategy. First of all, we are a member of the North Atlantic Treaty Organization (NATO), and there is some question as to why this strategy was put forward. Many people feel the reason we put it forward was to justify the greatly increased budget and particularly the greatly increased budget for the navy, but we didn't really consult our NATO allies when we did this. Yet the war plans of all the NATO countries require the ships of those countries to come under a NATO commander. That means that Canada is part of this maritime strategy. Denmark is part of this maritime strategy. Norway is part of this maritime strategy. We are all involved, and yet you did not have an opportunity to "buy off" on it.

Going back to the question of submarines, we anticipate that the battle will take place primarily in the Arctic. The Soviets operate their missile submarines in the Arctic. The United States will be in the Arctic attempting to sink Soviet missile submarines. Let me go back to the danger of that problem.

The United States says it is going to sink Soviet missile submarines at some time early in the war—five minutes after it starts, Secretary Lehman said. But we say, if we sink Soviet nuclear missile-firing submarines using conventional torpedoes, that this is not an act of nuclear war. That is like saying that it would not be an act of nuclear war if we were to destroy their land-based missiles using conventional bombs. The difficulty here is that the Soviets will consider it an act of nuclear war: they have said that if we sink their submarines, it will be an act of nuclear war. That is very significant. What is more significant? If we sink their submarines, and they become aware of this, and if they launch their nuclear weapons in order to use them before they lose them, we have obviously begun a nuclear war.

This could easily be the nuclear war by accident that people talk about. The attack nuclear-submarine commander who has a ticket to sink Soviet missile-firing submarines may even be a Canadian attack nuclear-submarine commander. The Soviets find out about it; the Soviets commence a nuclear war. This is almost nuclear war by accident, but the important thing is that if we have this strategic nuclear exchange we have lost the war. Both sides have lost the war. The world has lost the war. President Reagan has said it. Gorbachev said it. We cannot afford to have it.

There is another problem with this maritime strategy. It is not as significant as the last one, but it is very significant to Canadians. It is that as our fleet moves north we are clearly going to attack the Soviet Union by flying over the land masses of other countries. Both U.S. and NATO planes will be flying over Sweden, and U.S. and NATO cruise missiles will be flying over Sweden, for example. And yet Sweden was never asked, it was never consulted, about this maritime strategy. In fact, I've been in Sweden discussing this subject, and the people there say that one of the most embarrassing things for them in Sweden now is that the Soviets are asking them: "Why do your air defences only look this way? Why don't they also look the other way?" So there are many problems associated with the maritime strategy.

There is a reason not to be as concerned about the maritime strategy as what I have said might indicate: it is that it is a "strategy"; it is not a "plan." All of our NATO commands, all of our ships, all of our command centres have war plans in them. The war plans essentially are what you will do when you go to war. You have a national plan and you have a NATO war plan, depending on whether it is a national war or a NATO war. I am almost certain that those war plans were not changed one iota because of this maritime strategy. Many people question the military viability of this strategy. And the Secretary of the Navy who followed Secretary Lehman said it was nothing more than a strategy, so that it would be up to the wartime commanders to say whether it would be used in war, and to my mind it quite likely would not be used.

Let me now shift to the discussion of the U.S. concerns about nuclear-weapons-free zones. These, however, I shall describe as constraints on the U.S. nuclear arsenal since we view a nuclear-weapons-free zone as being an effort on the part of somebody else to put a constraint on our use of our nuclear arsenal.

As you heard yesterday, the United States signed the South American nuclear-free zone treaty. We signed it in 1969, under President Nixon. We did not sign the South Pacific nuclear-free zone treaty, which was done under President Reagan. Now what is the position of the United States on nuclear weapons? We have always said that our nuclear arsenal is for deterrence. In order to achieve this deterrence, the United States basically has two types of nuclear weapons today.

First, we have what we call strategic nuclear weapons. We have land-based strategic missiles—MX and Minuteman missiles, which are stationed in the western part of the United States. On these land-based missiles we have twenty-four hundred nuclear warheads. We have strategic nuclear weapons on our submarines. We have Trident and Poseidon submarines. The Poseidon submarines are in the Atlantic, and the Trident submarines are currently in the Pacific. We have fifty-five hundred warheads on submarines. The third leg of our strategic nuclear weapons is bombers. We have B-52 bombers; we have FB-111s in the United States; we have the new B-1 bomber. We have forty-nine hundred nuclear weapons on those bombers. The total arsenal of strategic nuclear weapons is approximately twelve thousand strategic nuclear weapons.

In addition to this arsenal, we have nuclear weapons in Europe. This is the second category. We have about forty-six hundred weapons in Europe, even after we have destroyed our intermediate nuclear forces, even after we have destroyed the ground-launched cruise missiles and the Pershing 2s. Those weapons in Europe are stationed on fighter-bombers, and we have short-range missiles, and we have some anti-aircraft weapons. But the Center for Defense Information calculates that eighteen hundred of those forty-six hundred weapons in Europe could be delivered on the Soviet Union directly from Europe. That is, they are intermediate nuclear forces, but the weapons that will deliver them could deliver them into the Soviet Union.

So we have about fifteen thousand nuclear weapons that we can deliver on the Soviet Union. Now that's deterrence. But why is it deterrence? When I came into the nuclear navy and the submarine force in the early 1960s, some people were talking about five hundred weapons as enough to do the job. Now we have fifteen thousand weapons. Without going into the destructive capability of these weapons, that, clearly, to my mind, is a deterrent force plus!

What I would like to say about this arsenal of weapons is that what the United States doesn't need to do is to add to that deterrent force. And what we do not need to do is to put ships armed with nuclear weapons in New Zealand. We don't have to put ships armed with nuclear weapons in Denmark or in Norway or in Canada. We do not have to have them. Can you see that that has absolutely no military significance to our deterrent force? It is of no importance. In addition, we also do not need to put nuclear weapons in the Arctic. We do not have nuclear weapons in the Arctic now, and there is no need for us ever to put them there. Look at that arsenal—fifteen thousand nuclear weapons. Why should we ever even contemplate putting nuclear weapons in the Arctic?

So the question is: Why would the United States oppose a South Pacific nuclear-free zone? Why didn't we sign that treaty? It makes no difference to us whether the South Pacific is a nuclear-free zone or not. Why wouldn't we sign a treaty making the Arctic a nuclear-free zone? We don't want to have a nuclear war—I don't think we want to have a nuclear war—and anything we can do to try to reduce the threat of nuclear war is good. Signing an agreement for a nuclear-free zone in the Arctic, I believe, would reduce the threat of nuclear war—as would a South Pacific nuclear-free zone. It is hard for me to understand why the United States opposes this.

Let me say that I understand that there are different paths to nuclear disarmament. Both President Reagan and President Gorbachev have indicated a desire for nuclear disarmament. To me a very good path is a nuclear-free-zone path. We have a nuclear-free zone in South America, we have a nuclear-free zone in the South Pacific, we've got a proposal for a nuclear-free zone in the Arctic. One of the things that these zones can do is to act as confidence-builders. As time goes by and these zones work, we are building our confidence.

The next logical step is a nuclear-free zone in Central Europe. Now that would hurt. That would hurt the United States. It would not be done without pain. We have forty-six hundred nuclear weapons there. We don't have any in the Arctic; we don't have any in any of these other places; so other nuclear-free zones would not be the least bit painful to us. Central Europe would be painful, but it is a logical step. If we are going to get rid of nuclear weapons, let's do it in Central Europe also. And then eventually we could work out ways to get rid of all nuclear weapons. That should be the goal. But the nuclear-free zones are good stepping stones, I believe.

Why does the United States not want to do this? It's hard for me to understand. I think there is a certain amount of feeling: If you like me, you like my nukes! We really just can't stand somebody being mad at us because we have nuclear weapons, particularly our allies. It really bothers us when our allies won't let us have nuclear weapons in their territories. Suppose that Canada had a policy of no nuclear weapons and that, if we said we had nuclear weapons on our ship, you would say that we can't come in; that would pain us. The same thing with the Danes. But actually it shouldn't pain us.

Let me say this. I was on nuclear-powered missile-firing submarines for ten years and captain for three years. During that period of time it was difficult to take those ships into foreign ports, not because of the nuclear missiles but because of the nuclear reactors. People didn't want to have nuclear reactors in their ports, and we understood that. I wasn't offended. There are a lot of ports in the United States that you can't take nuclear reactors into. I understood it, and my crew understood it. We weren't offended.

I do not understand why we feel this pain. Why is it? I can understand why you don't want nuclear weapons. I can understand why we have them today, and I could understand how some day we would like to get rid of them. But I also can understand why you do not want nuclear weapons in Canada, why they don't want them in Denmark, why they don't want them in Norway. And I don't understand why our government can't agree with me on this!

The other problem that I think stops us from joining the nuclear-free-zone movement is what we refer to in the United States as NIH. This does not stand for the National Institute of Health but for Not Invented Here. If it isn't an idea that we personally hold, we don't like somebody trying to tell us that this is the way we should go. And what we particularly don't like, what the American government particularly doesn't like, is having peace movements tell them what to do. When seventy percent of the American people supported a nuclear freeze, the American government resisted it. They resisted the objection of the European people to our deploying nuclear weapons. There is something about the American government that does not like to respond to the people. The interesting thing to me is how the Soviets have suddenly found that it is a pretty good thing to do, to respond to the people. I wish the American government would understand the same thing. Perhaps the best example was the Soviet eighteen-month moratorium on testing nuclear weapons. That was a response to the people. The American government should have followed, and we should have had a treaty banning nuclear-weapons testing.

I would like to spend just a couple of minutes talking about what Canada can do. First of all, you can and should lead the way. I think you should lead the way in nuclear-free zones. There were many statements made by various speakers this morning about why you can't do that. On one specific point I would like to point out why you can do it. One of the stated reasons that you cannot do it was that all of the Arctic nations should join at the same time if you are going to have a nuclear-free zone. I do not think this is necessary, and I think history indicates it is not necessary. One of the most successful treaties that we have had so far is the Nonproliferation Treaty that some 140 nations have signed today. They didn't do it all at once. In fact they continue to sign that treaty. So if you can just get one other country, get a few nations, then eventually pressure would build up. I think Finland would be willing to do it. Perhaps you could get Norway—I think Norway would come. I can just about assure you that Denmark would join. Get these nations, and then eventually pressure would build up. The United States would come. The Soviet Union would come. Maybe this is not going to happen next year but five years down the line or ten years down the line. Start it. Lead the way. I think Canada should lead the way.

There is another small point on which you can make your voice heard. It is in the same way that Denmark and New Zealand had their voices heard, but it would be far more significant to the United States if Canada did it. That is, as you heard the Soviet representative say this morning, start asking us whether we have nuclear weapons on our ships or not. This refusal to confirm or deny the presence of nuclear weapons is an outmoded, archaic policy.

Let me just repeat what I said before. I can understand why people don't want nuclear weapons, and I think our government should be able to do the same. If we come to you and say, "We would like to have this ship come and visit; it's got nuclear weapons on board," and you say no, we would say, "OK, we'll go somewhere else and visit." I was going to say that in the ten years that I was on nuclear submarines, when I was on the *Robert E. Lee* and the *John C. Calhoun* and captain of the *Simon Bolivar,* I never visited a port other than my home port. Not one port other than my home port did I visit in the ten years. And my morale was good! And the morale of my troops was good! We can do it. It is not an absolute necessity that we visit other ports. We should understand that. I think your voice can be heard if you just come out and ask, "Do you have nuclear weapons on your ships or don't you?" It is a very easy step.

Let me make one final pitch, if you don't mind. We have been talking a lot about nuclear-free zones and the reduction and perhaps the elimination of nuclear weapons. President Gorbachev has talked about this, and eventually President Reagan came to the same conclusion; that some day we should eliminate nuclear weapons. I think one of the things we should focus on is the fact that, if we do eliminate nuclear weapons, we cannot eliminate the technology. That point came out yesterday. So that if we had another war, there would be an immediate nuclear arms race, and I suggest that as soon as the weapons were built they would be used, because you would want to use them first if you could. So really what we are talking about is not the elimination of nuclear war. The enemy is war itself, and that is what we should focus on.

I would like to say one last thing. The question I have often been asked these days is: "Are you related to President Bush?" My answer is that I haven't decided yet!

CHAPTER 23
Question Session

CAPTAIN JAMES T. BUSH, ANDREY E.
GRANOVSKY, YURI V. KAZMIN AND
AMBASSADOR ALEXEI A. RODIONOV

Linda Hughes
Captain Bush, you talked about the reasons why the United States doesn't support a nuclear-free zone, and you cited some obviously perceptive ones. But I wonder, is it really because the United States doesn't want to talk about navies and fleets at all? They are happy to talk about land forces perhaps because that is where the Soviets have superiority, but they don't like to talk about the navies. Is that because, if they do talk about it, it opens up discussion of naval forces all over the world, and if we agreed on a nuclear-free zone in the Arctic, they would have to start talking about the Mediterranean, the Pacific and other theatres?

James T. Bush
I think that is exactly correct. The United States would not like to begin talks about naval forces because of the fact that we do clearly feel superior—although from time to time we will say the Soviets are superior in order to try to justify building more ships. In fact, the Soviets do have more ships than we do, but ours are far more capable. This is particularly true of aircraft carriers. I'm not trying to be chauvinistic here, but we have thirteen to fifteen aircraft carriers, depending on how many are in overhaul. The Soviets have no aircraft carriers. The aircraft carrier is a most significant naval weapon today. And as much as the U.S. navy talks about its weaknesses, it is in that one area that the United States is so far superior that we don't want to take a chance on limiting naval forces.

Gwynne Dyer
I want to direct to the Soviet delegation, and I will not be specific as to

whom, roughly the same question I asked the Canadian representatives this morning. In terms of creating a nuclear-free zone in the Arctic, one could, as Captain Bush suggested, simply go at it unilaterally with whomever else would go along. The question is: How might the Soviet Union go along at an early point? And would it be acceptable to the Soviet Union if a certain area along the Soviet coast were exempted, but the rest of the Arctic became a nuclear-free zone, with some provisions for access to the Atlantic of the sort I discussed this morning? Is that an interesting proposition?

Alexei A. Rodionov
First of all, I would like to assure you that the Soviet Union does not intend to exclude our Kola Peninsula and other points of our Arctic and the North from our efforts in the field of arms control. We are prepared now to discuss the Kola Peninsula as well as other parts of our North as a part of a general disarmament dialogue. For instance, a planned fifty percent reduction of strategic arms under a treaty being prepared between the Soviet Union and the United States will undoubtedly affect the missiles deployed there. I would like to ask Mr. Granovsky to speak more specifically on our approach to the idea of establishing a nuclear-free zone in the Arctic.

Andrey E. Granovsky
When I heard the proposition for the first time there was terror in my eyes, and I was about to say: Everything, but not this. But then the next minute I thought: Why not this? If an agreement for a nuclear-free zone in the Arctic is based on the preservation of both mutual security and the existing balance of forces on a minimum level, it is quite acceptable. The only thing is of course that details should be elaborated and that can be done only during the negotiations.

What is the situation? The situation is that practically all our sea-based nuclear forces are inside the Arctic, while the U.S. situation is absolutely different. So the inclusion of the Kola Peninsula and our northern territory in this zone means unilateral elimination of our sea-based nuclear missiles. But in the broader context things are happening now that one could not have imagined even five years ago. If there is a kind of compensation, for example, liquidation of both our sea-based nuclear forces and American sea-based nuclear forces, negotiation may be possible.

A fifty percent overall reduction would reduce to approximately fifty percent our nuclear forces inside the Arctic. But as Captain Bush was absolutely correct to say, we can now destroy each other forty or sixty or eighty times—I have not made the calculation. It has nothing to do with deterrence. We are actually against nuclear deterrence. We consider it dangerous and useless. But if deterrence exists and if there are people who believe in deterrence as a method of prevention of war—excuse my cynicism—it is sufficient to eliminate each other once, not forty or sixty times.

In this context, we can think about the radical reduction of nuclear forces. If there is a political will on both sides, anything is achievable during dialogue. Of course there will be difficult problems to discuss. For example, there is a legal problem if only parts of the territories of the Soviet Union and the United States, which are nuclear powers, are made nuclear-free. There are strategic aspects to discuss, but with good will and with patience everything can be achieved.

Ann Medina
I just have a brief question—though I must say I would love to have the Canadians come up here and discuss with Captain Bush why the Canadians cannot or have not taken some kind of an initiative. This morning we heard the Minister of Defence of Norway mention how cooperation in the development of natural resources could in fact provide a kind of deterrence, in terms of there being a mutual self-interest in keeping the Arctic, let's say, not vulnerable. I noticed again in the talks just prior to this, a lot of stress on possible "co-ventures" for the development of natural resources, such as oil and gas. I'm wondering if we are beginning to see mutual cooperation on developing our oil and gas resources in a way that possibly could be contrary to environmental interests.

Rodionov
We are interested in cooperation in the North and the Arctic. Eighty percent of the territory within the Arctic Circle belongs to our two countries, and this naturally leads to a mutual desire to develop productive cooperation between us. In accordance with the programs of 1984 and 1987, we established good cooperation in the fields of geology, oil and gas; the economic field and humanitarian components; and some legal aspects. Now there are very good opportunities for cooperation in the economic development of our bilateral relations, and one form of such cooperation could be the establishment of joint ventures. We have in this regard very good examples. We have signed an agreement with Foremost Limited from Calgary to establish a joint venture that will produce transportation equipment for extreme northern situations. We have very good prospects in such areas as oil and gas exploration. Mr. Kazmin will continue the reply to this question.

Yuri V. Kazmin
I understand the question to be how cooperation in non-military matters,

especially in the development of resources, may influence easing up the conflicts in the Arctic. I think that a good example of that is the Soviet-Norwegian cooperation on the management of fisheries in an area that politically is a conflict area, because we have not yet settled the delimitation problem of the economic zone and the continental shelf in the Barents Sea. However, in 1975 we established a joint committee to look after the fisheries there. It was a so-called grey zone agreement that operates in spite of the fact that we have not settled the problem of dividing the economic zone, which is a political problem. We managed to ease tension over the area through joint cooperation in the management of the resources and by good will. The Norwegians are cooperating with our efforts. I think the best thing is that cooperation in non-military areas leaves less room for conflicts in military areas.

Betty Brightwell (Voice of Women)
I live in Esquimalt on Vancouver Island, British Columbia, where it is expected that some of Canada's proposed fleet of nuclear submarines will port when not on patrol or under the Arctic ice. My concern is over the possibility of a spill of radioactive primary coolant. You are perhaps aware that British Columbia government agencies currently take water samples before, during and after visits of American nuclear hunter-killer submarines to Esquimalt to test whether or not there is any radioactive residue; so somebody is concerned about that. I would like to point out—and I am sure, Captain Bush, you know this—that in the 1970s in Apra Harbour in Guam a disabled American submarine leaked radiation to fifty times greater than the acceptable level onto local beaches; I want you to be able to assure me that an accident of this sort will not take place in the Arctic. Captain Bush, in light of the accident in Guam, I have some doubts as to the claim of the U.S. and Canadian submariners that there have been over three thousand safe reactor-years; I would like you to tell me if this is true or not. Incidentally, I have seen the directive advising the U.S. navy to deny or to play down any sort of a nuclear problem on its boats when dealing with the media, should there be a nuclear accident to its reactors. I would like you to comment on that too.

Bush
Thank you for the question. The nuclear reactors on the U.S. submarines, I believe, are fairly safe. They have a good record. There is no doubt there have been accidents that resulted in the discharge of radiation and have not been reported, but in general the record is good.

Let me talk about your first question, however, which is the possibility of primary coolant being discharged in your harbour. The United States used to discharge primary coolant directly into the harbour. We did it in

Holy Loch while I was on my first submarine, the *Robert E. Lee*. We stopped discharging primary coolant into the harbour because we felt that there was at least a residual radioactive problem there.

You are not going to be buying submarines from the United States; you are going to be buying them from the British or the French. Since this proposal has been made, I have been asking various people in Canada to find out whether the British and French submarines still discharge primary coolant into the harbour. I believe they do, but I have not found anybody here who can answer that question. Nobody in your government nor, as far as I can see, in your military has asked that question and received an answer. It seems to me that before you buy these submarines, you ought to find out whether they are going to be discharging primary coolant into the harbours.

I do know what the British and French say about discharging primary coolant. They say that they do not discharge radioactivity into the harbour. Let me tell you that when the United States was discharging primary coolant, we used to test it and we didn't think we were discharging radioactivity into the harbour, but we found out eventually we were. The British and French, when they say they are not discharging radioactivity, are not saying they are not discharging primary coolant, and I think that that is a very important distinction and one that Canada ought to find out before buying these ships.

Blair Davis (Yellowknife, Northwest Territories)
Captain Bush, in your speech a main topic for conversation was where nuclear warheads are *not* being stored, such as the Arctic, and many of the people were applauding that. But you said also that there are fifteen thousand nuclear warheads, which you are using as a deterrent against the Soviet Union. The megaton content of each of these warheads is probably very high, so I think that these warheads are more a form of aggression. Does Captain Bush honestly think that these warheads are a deterrent to war or a deterrent to global peace?

Bush
It is very difficult to deal with questions on deterrence. My standard answer is that if it is true that the U.S. nuclear arsenal has prevented the Soviet Union from attacking the United States using nuclear weapons, then it is also true that the Chinese nuclear arsenal has prevented the Soviet Union from attacking the Chinese using nuclear weapons. The Chinese have six long-range missiles. They have two submarines with fourteen missiles on each submarine. I believe that deterrence is a difficult subject to discuss, and I would go along with my friend from the Soviet Union who says he does not believe that nuclear deterrent forces are necessary or

actually represent deterrence, certainly not forces this large. But there are many people who do. What we have to do is to eliminate nuclear weapons, and we have to do it together; that is the proposal. Eventually we will pick up the British, the French and the Chinese also. The Chinese have already said they would join us. There would be a great deal of pressure on the British, and I am sure they would join us. I'm not sure what the French would do.

Eric Tollefson (Canadian Pugwash Group)
Captain Bush, we in Alberta are very concerned about testing the cruise missile. We look upon it as a weapon capable of supporting a first strike, and one that could lead to nuclear winter, the deaths of millions of people and perhaps the end of our civilization as we know it. Is this an area in which Canada could take the initiative and say to the United States, "Look, we are sorry; we no longer want to test your cruise missiles"?

Bush
Very definitely; this is clearly a point that is very important. The testing of cruise missiles in Canada is very important to the United States. The topography of Canada is similar to the topography of the Soviet Union. But we don't need cruise missiles either! Let me just say one other thing about our great democracies, as we call them: it was very difficult for me to believe that this decision to test cruise missiles in Canada never even was subjected to a parliamentary vote.

Hazel McLarty (Edson, Alberta)
First of all, I would like to say that I think military concerns are no longer the issue here. I think we have come full circle to being concerned about life itself. At a recent symposium of environmentalists, one of the scientists got up and said, "I think eventually the women of the world will have to take a stand if we want anything to be done." In that regard, I would like to ask all of the representatives here, but particularly the people from the Soviet Union, how are the women involved in these decisions in your country?

Adrienne Clarkson
A dilemma! A male dilemma! Mr. Kazmin is leaping into the lion's mouth.

Granovsky
It turned out to be the most difficult question for us!

Kazmin
I will say that women in the Soviet Union are involved in the decision-making to the same extent that men are.

Clarkson
That's an answer. Thank you very much.

McLarty
I'm sorry to hear that.

Harriet Critchley

CHAPTER 24
Canadian Policy Options for the Arctic

HARRIET CRITCHLEY

I have been asked to look at Canada's policy options and our choices for peace and security from a strategic perspective on the Arctic. By "strategic" I mean the broad view, identifying trends over longer periods of time. From this perspective one major trend stands out very clearly: the strategic significance of the Arctic (and by that I mean the whole Arctic region) has increased rather suddenly and dramatically over the past twenty years.

Four factors are involved in this trend. We have been discussing them over the past day and a half. I might have slightly different words for them or put them in a slightly different order, but they are what we have been talking about.

First of all, there have been changes in the international Law of the Sea. Professor Pharand's excellent presentation yesterday mentioned

Harriet Critchley is an associate professor in the Department of Political Science and program director of the Strategic Studies Program at the University of Calgary. She holds a Ph.D. from Columbia University. She has published extensively on Canadian defence and arms control policies, circumpolar relations and political development in the Canadian North. She acted as a consultant to and member of the Canadian delegation to all three United Nations Special Sessions on Disarmament and was also a member of the Consultative Group on Disarmament and Arms Control Affairs. She has served on the task force on Review of Unification of the Canadian Armed Forces, the executive committee of the Canadian Military Colleges Advisory Board and the board of the Canadian Institute for International Peace and Security.

some of them as they affect Canada's Arctic waters, but they affect other Arctic states as well. They are the establishment of the two hundred mile zones and the boundaries between them, the extension of the territorial seas from three to twelve miles, and the rules governing navigation in ice-covered and ice-infested waters.

Second, there have been changes in the economic values of the renewable and nonrenewable resources. Again this was mentioned by Mr. Weick and Mr. Berger yesterday, and again it affects the other Arctic states, all of which have potential or discovered oil and gas deposits. Many of them have potential and discovered deposits of other minerals, mainly metals, and many have important renewable resources, for example, fish.

The third factor has been northern political development. Over the past day and a half we have discussed various facets of the increased political participation in, and the increased awareness of, national and international issues by residents of the northern parts of the Arctic states. We have talked about their increased political activity and their increased role in decision-making and planning for the economic development of northern areas. We have talked about their increased concern for the environment in general and in particular for the environmental effects on the Arctic of economic development in the Arctic and elsewhere.

There has also been increased enunciation of northerners' values and priorities for this general political development. A number of the speakers have mentioned this, but again I would like to draw to your attention that it affects most, if not all, the Arctic states.

The fourth factor is changes in the superpowers' military and strategic doctrines, which have been brought on by changes in military technology. The result, as we have heard discussed repeatedly at this inquiry, has been increased military activity in and near the Arctic. Again this factor affects all of the Arctic littoral states directly.

Because of these four factors taken together, the Arctic states and others are paying much more attention to the region. This is for a variety of reasons. The region is more important, it is more valuable, and it is of more concern than it was twenty years ago. This kind of change in the strategic significance of the Arctic generates new issues for bilateral and multilateral cooperation, and we have heard examples of those. It also generates new issues for bilateral or multilateral conflict, and here I would give you the example of the conflicts between a number of Arctic states over their two hundred mile zone boundaries. This change in the strategic significance of the Arctic happens to coincide with, I would argue, another set of changes that have been occurring in Canada over the same past two decades. Again taking the broad view, there are two major factors in these changes.

First, there has been an increase in Canadian self-discovery and self-confidence as a nation. Not so long ago the popular question was: What is a Canadian? And the answer was often couched in terms of what we are not—that is, we are not Americans—rather than in terms of what we are. But that question is rarely asked now; have you noticed? I think that there is a significant Arctic component in this Canadian self-discovery and in the national identity of Canada and Canadians.

Second, accompanying this change in self-discovery and self-confidence has been an increasing realization of Canada's place in the world. This is different from the earlier Canadian perception of being isolated from and insulated from the world, off in our little northern corner at the top of the Mercator projection, where there is nothing north of us but the wall to which the map is attached. We are now seeing ourselves as a trading nation and a maritime nation with important interests across the Atlantic, across the Pacific and around the Arctic. In short we are realizing that these three oceans link us to the world rather than separate us from it.

I suggest that these two sets of changes, that is, the increasing strategic significance of the Arctic and the increasing realization of a distinct Canadian identity and Canadian place in the world, insofar as they apply to the Arctic, can be taken as the context for this inquiry. Otherwise why are we all here? I suggest that we recognize what I have called the changes in the strategic significance of the Arctic and that we are concerned with what we as Canadians can or should do about it. What are our policy options in the Arctic? What are our choices for peace and security?

In addressing this central question, two things should be kept in mind. The first is that the sources for the change in the strategic significance of the Arctic are often mainly outside Canada and are not of the making of individual Canadians or Canadian governments. For example, changes in the international Law of the Sea are a product of both complex, multilateral negotiations and the practice of many individual states. Changes in the economic value of the resources are often the result of the forces of world supply and demand and of the decisions of private and state enterprises based on their own investment and profitability requirements. They are not necessarily the results of our national interests. We do have the main influence on political development in Canada, which is of course the result of our own political, economic and social values and processes and priorities. There are, however, other Arctic states that have their own values and processes and priorities, and each state is the main influence over its own political development. Changes in superpowers' military and strategic doctrines are mainly the province of the superpowers and are a function of their interaction over time.

Because our influence over the sources of change may be limited, our options may also be limited—limited in the sense that we are reacting to change that has its source elsewhere and whose course we have some difficulty in predicting over even the medium term, not to mention the long term. For example, I draw your attention to the various predictions we have made over the past two decades of where world oil prices were going and to our national plans in Canada based on those predictions. About the only thing you can say with any certainty is that the predictions were always wrong! While changes in the strategic significance of the Arctic have a direct bearing on peace and security in Canada, they also have a bearing on the peace and security of other Arctic states as well as some non-Arctic states, which those states evaluate from their own perspectives.

The second thing that I think we should keep in mind in addressing the central questions of policy and of peace and security is that, while there is a general worldwide agreement on the meaning of the word "peace," the same may not be true at all for the word "security." In fact, looking only at those states bordering directly on the Arctic, I suggest that, for some, security is seen primarily as a matter of national defence; examples are the United States and the Soviet Union. For others, security is seen primarily as a matter of social development and fiscal viability; an example is Greenland, as a part of Denmark. And finally, for others, security is seen primarily as a matter of sovereignty assertion very broadly defined; Canada and Norway are examples.

In fact, within Canada (and within some other states) we are engaging in a debate precisely about the meaning of the word "security." Regardless of the particular perception of the term, the different mixes of interests among Arctic nations create conflicts of views or, at the very least, difficulties in negotiating cooperative arrangements. In spite of those difficulties, however, cooperative efforts to resolve common concerns and problems have emerged.

In short, we should keep in mind that when discussing Canada's options for policies on peace and security in the Arctic we are trying to assert Canadian sovereignty in a region where change is occurring rapidly, where there are many different perspectives and interests at work, and where we often have limited influence over the sources of change.

I think we also have to be clear on the use of the term "sovereignty." Strictly speaking, it is a term of international law that refers to the jurisdiction of a state's laws over a defined territory, including land, water and air. There is, however, a broader implication to that strict legal definition because our ideas, our values and our institutions are embodied in our laws and regulations. Included in Canadian values are the high priorities we attach to tolerance, cooperation, negotiation and multilateralism. In asserting sovereignty, we are asserting those values, among others, and exercising those values in our dealings with other states. In short, we are acting as one sovereign state in an international system made up of sovereign states.

Let me give one example of Canada's assertion of values. As Canadians, we attach a very high value to pollution prevention and that value is increasing—in fact, issues created by environmental pollution may be the first priority for Canadians today. With respect to the Arctic, since the early 1970s the government of Canada has defined the prevention of pollution from marine sources as vital to the security of the state. That was provided for by the Arctic Waters Pollution Prevention Act of 1970, promulgated in 1972. At that time we were one of only a few sovereign states, if not the only one, to define the security of the state that way.

Thus in 1985 when we drew straight base lines around the Canadian Arctic Archipelago, declared the waters within to be internal waters and by this means asserted Canadian sovereignty in the Arctic, we did not do so for the purpose of excluding other states. Rather our purpose was to make sure that vessels obey Canadian laws and regulations so as to ensure environmentally safe passage. Achieving this purpose promotes Canadian security. Our policy suggestions and actions with other states with respect to air-source pollution, ozone-layer depletion, the Polar Class 8 icebreaker acquisition, scientific and technical cooperation, and wildlife preservation are all to be seen as directed toward the same goal.

Canada also promotes its security by Canadian defence activities in the Arctic. But first we should note—and it has been mentioned at this inquiry—that Canadian defence activities are not very great in relation to the size of our Arctic region. We have some sixteen overflights a year by unarmed patrol aircraft. We have created a native, predominantly Inuit, paramilitary force, the Rangers, who live in the communities of the Arctic. We have occasional small-scale army exercises. We have a radar early warning system for aircraft. We have some plans that were contained in the White Paper on defence; for example, for five to six airstrips to enable us to land our fighter aircraft for the purposes of refuelling them; for an underwater detection grid, in a way much like an underwater version of our early warning radar line; and possibly an occasional pass through our waters by one of our submarines, if the government decides to go ahead with its plan for the necessary submarine purchases.

You should note that Canada has no combat aircraft based in the North. Canada has no ground combat units based in the North. Canada has no missile installations anywhere. It has one small headquarters located in Yellowknife, in the southern part of the Northwest Territories. That headquarters has about seventy people. It is so small that when Cosmos 954 crashed in the Northwest Territories, the Canadian rescue and decontamination effort was organized out of Ottawa and Winnipeg, and not out of Yellowknife. None of the other Arctic littoral states with the exception of

Greenland has so few military forces in its Arctic regions.

Now this may be controversial, but I suggest that another way of looking at these activities and plans is that they are assertions of Canadian sovereignty. We claim jurisdiction by surveying the territory over which we claim sovereignty. We do so to ensure the observation of Canadian laws, laws that reflect our values. An example is multilateralism, a value we practise by honouring the promises made to friends, allies and others for cooperative coverage of our offshore air and sea zones. By ensuring that Canadian laws are observed, we ensure that activities on, over and under our territory occur only with the knowledge and the lawful agreement of our government, that is, by negotiation, whether it is domestic or international negotiation, in order to exert some control over the activities of other states.

I have used these two examples intentionally. The government's response to the House-Senate special joint committee was an agreement to four themes for Canadian Arctic policy. The themes were sovereignty, defence, use of the Northwest Passage and circumpolar cooperation. I have argued here today that all of these actually constitute sovereignty assertion in the broadest and, I would like to suggest, the best sense of that term. I therefore suggest that the theme of sovereignty assertion should be our guide in thinking about discrete Canadian policies for peace and security in the Arctic. It may just be our most useful guide when we attempt to carry out the very delicate and difficult task of promoting our values, such as tolerance, cooperation, negotiation and multilateralism, for a region where change is occurring rapidly, where there are many different perspectives and interests at work, and where we often have only some limited influence over the sources of change and over the perceptions of others.

CHAPTER 25
Take It from the Top: A Proposal for a Nuclear-weapons-free Arctic

STEVE SHALLHORN

As you may know, Greenpeace is an organization that was founded in Canada in 1971 and has since grown to be active in about twenty countries. By the end of 1989 we expect to be active in all of the Arctic Rim countries.

The nuclear arms race has spiralled to the top of the globe. As Canadians we have to face the fact that we have allowed our country to contribute to that. It may be hoped that this True North Strong and Free inquiry will act as a catalyst to first stop and then reverse that arms spiral.

I believe that the Mulroney government's White Paper on defence policy and its Cold War rhetoric were outdated even before the paper was released. Certainly changes in the international climate in the past eighteen months have rendered it obsolete and useless as a policy document. In particular, the decision to acquire nuclear-powered submarines is a clear step in the wrong direction. There have been others. Coupled with a dearth of any diplomatic initiative even to reduce tensions, let alone to reduce the level of Arctic military activity, Canada's position, I believe, is offensive to Canadians who believe that we should undertake an active foreign policy to pry the superpowers apart and to preserve the Arctic.

Canadian membership in the North Atlantic Treaty Organization (NATO) and the North American Aerospace Defence Command (NORAD) has skewed our diplomatic and defence thinking. It has been almost half a century since the outbreak of a world war, yet our thinking is still locked into a Second World War mentality. A Canadian military presence in Europe is not a prerequisite of NATO membership. The alliance was formed without such a presence; troops and aircraft were not agreed to as part of that membership. Yet we spend money to maintain a

Take It from the Top: A Proposal for a Nuclear-weapons-free Arctic

Steve Shallhorn

token military force of no particular military value on another continent as a way of trying to have some political leverage in Europe. Yet at the same time we allow a superpower to set up its military apparatus on our territory and to subsidize our own national defence system. We lose political leverage in our own backyard when we do that.

Canada's legitimate peace and security concerns in the Arctic are very closely tied to the regional military activities of the two superpowers, the United States in particular. In fact, Canada's own military presence in the Arctic has twice, in the 1950s and the 1980s, been initiated at the behest of one of the superpowers, the United States.[1] Each time, this military presence came when American domestic politics were dominated by a mass phobia of the Soviet Union.

I now turn to a discussion of military and naval activity in the Arctic. In this discussion, the Arctic will be defined as north of the 60th parallel. Countries with territories north of 60 include Canada, Denmark (which includes Greenland), Iceland, the Faroe Islands, Norway, Sweden, Finland, the Soviet Union and the United States.

In recent years both the United States and the Soviet Union have substantially increased their military activity north of 60. The most dramatic increase has come in the physical presence of naval forces, particularly nuclear-powered submarines of the two protagonists.[2]

[Editor's note: Passages in italics were in the written speech but were not delivered orally because of time constraints.]

Most Arctic naval activity is taking place outside the Canadian Archipelago, in the waters east and north of Norway and the Soviet Kola Peninsula (Norwegian and Barents Seas respectively). Soviet naval bases on the Kola Peninsula are the only ones that give them access to the open waters of the Atlantic Ocean. Three factors have led to a situation where American and Soviet submarines play a continuous and deadly cat-and-mouse game in Arctic waters. First, the U.S. and NATO navies created a formidable barrier of anti-submarine defences in the approaches to the Atlantic, from Greenland to Iceland to the United Kingdom (known as the

Steve Shallhorn is coordinator of the Nuclear Free Seas Campaign for the worldwide Greenpeace organization. Mr. Shallhorn received two bachelor's degrees from McMaster University, which gave him the Honour M Award for contributions to the university. After coordinating the Greenpeace disarmament program in Canada for two years, he joined the office in Washington, D.C. He has been active in the disarmament movement since the early 1980s and his special interests include Arctic disarmament, Canada–U.S. defence relations, and naval disarmament. He has served on the Consultative Group on Disarmament and Arms Control Affairs in Ottawa.

GIUK gap). *Consequently, Soviet ballistic-missile submarines (SSBNs) could not move close enough to the United States to use their missiles without a high risk of detection or, during war, destruction. Second, by the mid-1970s the Soviets developed submarine-launched ballistic missiles (SLBMs) with enough range that they did not have to get through the GIUK gap in order to be within range of American targets. The Soviets could now use the deep waters of the ice-covered Arctic Basin without fear of surface warships and with some protection by their long-range aircraft. Finally, the Americans began to articulate a new, aggressive forward maritime strategy to go after Soviet ballistic-missile submarines in Arctic waters with nuclear-powered attack submarines (SSNs) and to move aircraft carriers into the Norwegian Sea, close enough to launch nuclear-armed aircraft at Soviet targets.*

What is publicly known of naval activity within the Canadian Arctic Archipelago suggests that it is entirely that of Western countries. In the past few years, the Canadian military has told the House of Commons Standing Committee on National Defence that no Soviet submarines have been in Canadian Arctic waters and that, if they had been, Canada would know about them. American and British attack submarines have recently rendezvoused at the North Pole, but the two governments have refused to disclose the routes taken by these vessels.

The United States has undertaken a number of high-profile surfacings through the ice. In 1986, the Secretary of the Navy, John Lehman, joined three American submarines that had pushed their way through the ice at the North Pole. A map on page 52 of the White Paper on defence uses red arrows to suggest the intended routes of Soviet submarines transiting the Arctic to the Atlantic or Pacific. In fact, the arrows are pointing in the wrong direction: the Arctic is more likely to be used by American attack submarines surging their way to Soviet submarine bastions in the Barents Sea. Despite the words of the White Paper and of former Minister of Defence Perrin Beatty, sea lines of communication interdiction is a low priority for the Soviet submarine force, and this has been acknowledged repeatedly by the U.S. navy.

Since 1986 the operating tempo of the Soviet navy has been in decline. In February 1989 the U.S. Director of Naval Intelligence, Rear-Admiral Thomas A. Brooks, in testimony to the Seapower, Strategic and Critical Materials Subcommittee of the House Armed Services Committee on Intelligence Issues, which was chaired by the present Secretary of Defense of the United States, said, "In 1988 the Soviets scrapped or otherwise took out of service more ships than in any other year in recent history. In 1988 Moscow also began selling major combatants for scrap on the world market." Rear-Admiral Brooks went on to say that he expects overall numbers in the Soviet navy to decrease over the next five years. (Since unveiling their plan to buy nuclear-powered submarines, Canadian politicians and defence officials have taken to raising the spectre of over three hundred Soviet submarines lurking beneath the depths presumably waiting each in turn for their chance to sneak into Canadian waters.)

Rear-Admiral Brooks also pointed out that the Soviets have only about thirty available nuclear-powered attack submarines that are first line anti-submarine-warfare vessels, most of which are Victor III class submarines, "whose combat capabilities are less than the *Akula, Sierra* [Soviet SSNs] or modern U.S. SSNs."

So why is the Canadian government proceeding with its ridiculous plan to buy nuclear-powered submarines? The government's justification has tended to shift according to who the audience is. The public justification is that nuclear-powered submarines are necessary to protect Canada's sovereignty in the Arctic. The justification more often used in the so-called expert debate is that submarines are necessary to protect sea lines of communication during conventional war. Neither stands up to close scrutiny.

Perrin Beatty and the Department of National Defence went to great lengths in 1987 to convince Canadians that nuclear-powered submarines would protect Canadian interests in the Arctic. Five weeks before Canadians were given the report, a deliberate leak to American newspapers emphasizing the Arctic role resulted in both *The New York Times* and *The Washington Post* carrying front page stories on the plan the same day. Similar stories began to appear in the Canadian press. The idea that nuclear-powered submarines were intended for use in the Arctic was firmly planted with the media and with the public. After the *Polar Sea* debacle and American submarine surfacings at the North Pole, the media and the Canadian public were only too happy to accept the rationale.

By the time the election rolled around, however, nary a word about nuclear submarines was uttered by the Mulroney government during the campaign. After all the fanfare when the White Paper was released, the Conservative party neglected to make its defence program a part of the election platform. The Canada-should-do-more Mulroney-Beatty plan faltered because the internal logic behind the major initiative—nuclear-powered submarines—was flawed and contradictory. But most important, the overall goal in defence buildup and nuclear proliferation into Canada was out of line with what was going on in the rest of the world. That is why the Canadian public has moderated its support for the White Paper and for nuclear-powered submarines.

The Mulroney government has failed to come up with a credible political, or even military, rationale that the public and media will rally behind. It has been furiously scrambling to come up with the right formula to get the editorial writers and the cartoonists to stop picking on its idea. We are about to see the third or fourth round of that scramble, so we can expect to hear that the plan has now been reduced from ten or twelve submarines to perhaps five or six, and statements are being made in *The Financial Post* that the program needs to be resold. Apparently, consistent polling showing that seventy-one percent of Canadians are opposed to nuclear submarines is not big enough or large enough writing on the wall for the government.

Closer scrutiny shows that the government was attempting to use public opinion in favour of decisive action over American encroachment in the Arctic to justify a weapons system intended for something else. First and foremost, under NATO Canada cannot mount independent submarine patrols in the Atlantic and hence the Arctic. Under a process called "water management," all NATO submarine patrols are closely coordinated in grids by a NATO admiral acting as Supreme Allied Commander Atlantic (SACLANT), who is always coincidentally an American admiral. The rationale is straightforward enough from a military perspective. In a crisis, NATO commanders would need to know in a hurry if a sonar contact is one of their own or a potential adversary.[3] But in the process, Canada loses the capacity for unilateral action with the submarines.

Second, Canada does not plan to buy its own system to communicate with its nuclear submarines, which is a highly technical and expensive undertaking. Instead it plans to lease an existing American communication system, which will mean using American-supplied codes. It will not be possible for Canada to communicate with its submarines without the United States knowing the content of the messages and the location of the submarines.

That is why when Perrin Beatty finally stood up in the House of Commons after the pre–White Paper media blitz he had a different story. All of a sudden, Arctic duty was an option, a secondary role after protecting sea lines of communication in the Atlantic and the Pacific.

This blatant bait and switch was accompanied by a little bit of fibbing about the cost. There was a new term, the "sail-away" cost; it has a sort of drive-away feeling to it, like when you buy a new car. The submarines would cost $350 million for the French copies; $500 million for the British. It was dutifully reported at the time that the cost was about the same as the frigates that Canada was building. It wasn't until the camera lights had faded a few weeks later that the Minister appeared before the House of Commons defence committee and said that the program cost was moved up to $8 billion—and of course very few people believe that the submarine program will come in at $8 billion.

What Arctic disarmament initiatives should Canada undertake? Strategic naval activity is the greatest impediment to the creation of a multilateral nuclear disarmament regime in the Arctic. Submarine activity is by its nature global, covert, and difficult and expensive to monitor. The importance placed on ballistic-missile submarines by the superpowers and the geographical fact of the importance of the Kola Peninsula to the Soviet Union's naval forces are such that a demilitarized or denuclearized Arctic is closely connected to the disappearance of nuclear-powered submarines and of all naval nuclear weapons.

The one way to make sure that no submarines from any countries steal through the Canadian Arctic is to ban the only technology that allows them to operate in the first place, nuclear propulsion. A total ban on the use of nuclear power for naval vessels would be easily verifiable, would benefit all parties, and would remove the worry of disposal of nuclear waste created by nuclear submarines. Such a ban would create a much needed reorientation of navies from essentially offensive to essentially defensive postures.

Worldwide there are 544 nuclear reactors used for naval propulsion, more than are used for electric power generation. Nuclear reactors in fact were developed in the United States by the navy for use in submarines to provide for long-distance, high-speed underwater travel, which would allow extended covert operation along an adversary's coast. They were later adapted to generate electricity for utility companies. Reactors have so far allowed five countries to operate mobile, underwater missile silos in the form of ballistic-missile submarines (SSBNs).

A ban on nuclear propulsion could be incorporated with, but should not be tied to, SSBN sanctuary zones. As usually proposed, such sanctuaries would be close to a home coast where adversaries would be prohibited from operating submarines, ships or aircraft, or other sensors. This would allow the nuclear powers some confidence that their submarine-launched ballistic missiles would be available when they wished to launch them—if that is security. While sanctuaries are better than nothing in that some limitations are better than the current nuclear free-for-all, they do nothing to limit technologies available to nuclear powers, which means that they can instantly be revoked by one or all sides. We should keep in mind that it might not be too long before more countries build submarine-launched ballistic missiles and the submarines to carry them.

Banning the use of naval nuclear propulsion would eliminate the production of nuclear waste generated by the reactors, and it would eliminate the need to dispose of the radioactive reactors and hull portions of radioactive submarines. When they first built these things, not much energy was put into thinking about what to do with them when they were decommissioned. Early nuclear-powered submarines were filled with concrete and dumped at sea, but in 1983 the London Dumping Convention[4] prohibited ocean dumping of radioactive waste, which included the submarines and the reactors. Currently there are twelve nuclear submarines awaiting disposal in Bremerton, Washington, alone, with the current method of disposal being burial. At Rosyth in Scotland, the first British nuclear-powered submarine, HMS *Dreadnought,* is sitting on a great podium inside the shipyard while Britain tries to figure out what to do with it.[5]

Banning the use of naval nuclear propulsion is a technical solution to a technical problem that does not require revision of maritime law or restriction of freedom of the seas, which sanctuary zones or nuclear-weapons-free zones often require. The cause of nuclear nonproliferation would be served by banning nuclear propulsion. Already India has leased a nuclear-powered submarine from the Soviet Union, and Pakistan has reportedly asked the United States for one as well.

A ban on nuclear propulsion should include on-board verification of the propulsion system. To avoid emerging technologies allowing similar capabilities, these emerging technologies should be banned from use in submarines as well. Provision could be made for periodic surfacing and exposure to surveillance by satellites or aircraft to ensure compliance by all submarines.

Another disarmament initiative concerning naval vessels, and one with clear implications for Canada, would be to ensure that sea-launched cruise missiles (SLCMs) are included in a Strategic Arms Reduction Talks (START) agreement. Right now these missiles are the main sticking point between the United States and the Soviet Union, preventing agreement at the START talks. These relatively new weapons are carried on surface ships and submarines. They often carry nuclear weapons and can strike at each other's home territories. Both the Soviet Union and the United States are outfitting large numbers of surface ships and submarines to launch these weapons. Currently the United States does not want to include them in a START agreement, but the Soviet Union does.

It is difficult to understand why the United States wants to keep SLCMs out of a START agreement. Earlier this year a report titled "Deterring Through the Turn of the Century" by the high-level Discussion Group on Strategic Policy indicated a possible shift in the position of the United States. The nine-member group included Brent Scowcroft, President Bush's National Security Advisor, and powerful defence politicians such as Sam Nunn, John Werner and Les Aspin. The report stated, "With regard to nuclear-armed SLCMs, there is a basis for concluding that parallel U.S. and Soviet deployments would yield the Soviets a net advantage." In other words, it is in the U.S. interest to ban SLCMs.

I heard earlier today contradictory statements on Canada's position on SLCMs. General Huddleston mentioned at one time that if ballistic missiles are included in the START agreement—and presumably he meant that if SLCMs are left out—SLCMs will all of a sudden become very important. That is correct. But at the time he didn't mention what Canada's position was. Later on, in a response to a question, he did say that Canada is pushing to have SLCMs included in the START agreements, and I think it is very important that Canada push hard—vocally, publicly and privately—to ensure that this happens.

It is definitely in Canada's interest to ban these weapons. The Department of National Defence periodically talks of the grave threat posed by Soviet SLCMs. If they are prohibited, then Canada will no longer be vulnerable. Not incidentally, Canada will have less need for northern radars and fighter aircraft.

A final suggestion is to ban all tactical nuclear weapons at sea. Although this would not be limited to the Arctic, such a ban would be a good step towards elimination of all nuclear weapons at sea, including submarine-launched ballistic missiles. Worldwide, as of December 1987, there were over sixty-five hundred naval tactical weapons[6] and another two thousand ballistic missiles, strategic weapons, at sea. These weapons are carried on board ships into countless harbours—many of them in the centres of large cities. They are carried on ships that brush up against opponents, they have fires, and they sink.[7]

Regional nuclear-weapons-free zones are initiated presumably because there is a greater chance of success than applying them globally. In theory, this should hold true of the Arctic because of the relatively few number of countries involved, and except for the two superpowers, these countries have renounced deployment of land-based nuclear weapons during peacetime.

An Arctic nuclear-weapons-free zone should be a prime objective of Canadian foreign policy. In this case the Soviet Union would have to remove land-based nuclear weapons from the Kola Peninsula and northern portions of the Soviet Union, and the United States would have to remove them from Alaska. Removing nuclear naval weapons from north of 60 would be difficult because of limited Soviet access to the Atlantic. An Arctic nuclear-weapons-free zone combined with a ban on naval tactical nuclear weapons, a ban on nuclear propulsion, and ballistic-missile submarine sanctuaries would provide comprehensive protection of the Arctic. Of course, a complete ban on all nuclear weapons at sea would remove all difficulties presented by naval forces.

A nuclear-weapons-free Arctic proposal would have the advantage of separating the superpowers' nuclear weapons from each other. It could also have a positive effect on the current proposals for a Nordic nuclear-weapons-free zone in northern Europe and a nuclear-weapons-free Baltic. In both cases, the Nordic countries have been reluctant to agree to a proposal that would keep in place the heavy concentration of nuclear weapons on the Kola Peninsula, and the Soviet Union has been reluctant to remove weapons from a region that would not require the United States to give up some of its nuclear deployment areas.

This initiative could be implemented by the remaining seven Arctic countries themselves becoming nuclear-weapons-free zones while leaving the superpowers' arsenal intact. A few years ago this proposal seemed more likely than a comprehensive Arctic nuclear-weapons-free zone. Progress in the whole range of relations between the Soviet Union and the United States may have conspired to make a limited zone not worth pursuing.

Either a limited or a comprehensive Arctic nuclear-weapons-free zone for Canada would mean no port visits by nuclear-armed vessels. It would mean abandoning cruise-missile testing and low-level flight training of nuclear-armed bombers in Canadian airspace and would mean ensuring that underwater sonar systems are used for domestic surveillance only. Canada might agree with Denmark to build a joint system to monitor traffic between Canada and Greenland. But such a joint venture would have to differ significantly from the current cooperation between the two countries and the United States. Denmark is allowing the United States to do sophisticated anti-submarine research on ice islands on the northeast coast of Greenland, called Station North. The Danish press has stated that Canadian Forces Base Alert on Ellesmere Island is used as a staging point for supplies and personnel destined for Station North.

The problem with the Mulroney government's position on foreign policy is that it reflects how this government has become what we used to think the Kremlin was: old thinking—rigid—behind the times. Mulroney's idea of foreign policy has been free trade with the United States. The government has not seemed to notice that the Intermediate-range Nuclear Forces Treaty, the withdrawal of Soviet troops from Afghanistan, and Reagan's tour of Red Square have happened. Mulroney is one of the few NATO leaders who has not had a summit with Gorbachev.

Recent reports that the government will soon announce its intention to go ahead and build nuclear submarines but in fewer numbers indicates that there is still no new thinking in Ottawa. Committing Canada to a twenty-five-year nuclear submarine program at this time is simply absurd. Things are changing so quickly that I believe it would be a serious mistake for Canada to launch itself into such a long-term nuclear program.

If there is anyone with courage in the departments of National Defence or External Affairs he should make the point to the government that times have changed. The White Paper on defence needs a complete revision, not just a new cover. Plans for nuclear submarines and increased expenditures in Europe are the most obvious pieces that need to be dropped. But most importantly, Canada must actively pursue its own disarmament policy interests. Perhaps this inquiry and its participants can help those officials find their way to a strong but peaceful Canadian voice to wind down the arms race. We can start in the Arctic by taking it from the top.

Notes

1. Reference is to the Distant Early Warning (DEW) Line in the 1950s and the North Warning System (NWS) in the 1980s. (Editor)

2. There have also been significant upgrades in air attack and air defence forces, with the prospect of some sort of space-based defence in the future.

3. By assigning each submarine a designated area, they can avoid mistaken identity. While all Allied navies are required to coordinate their patrols with SACLANT, they are not told where other submarines are. That means that the United States will know the position of Canadian submarines, but Canada will not know the position of American submarines.

4. Officially the International Convention on the Prevention of Marine Pollution by Dumping of Wastes and Other Matters, it was established in London in 1972 by eighty countries.

5. Canada has not said how it would dispose of its decommissioned nuclear-powered submarines.

6. There were 6,584 naval tactical nuclear weapons, carried by 231 submarines, 645 surface ships and 3,198 aircraft. They include coastal missiles, naval artillery, free-fall bombs, anti-aircraft and anti-submarine weapons and SLCMs (which can also be included as strategic weapons).

7. Approximately 130 nuclear weapons are currently at the bottom of the oceans.

CHAPTER 26
Question Session

HARRIET CRITCHLEY AND STEVE SHALLHORN

Marie Zarowny (Roman Catholic Diocese of the Mackenzie, Yellowknife, Northwest Territories)
My question was originally for Mr. Roche, but part of it applies to Dr. Critchley's presentation, so I would like to address it to Dr. Critchley. You seemed to be justifying the increasing militarization that we are experiencing in the North as part of a broader definition of sovereignty that you gave us; and it seemed to me you were somewhat minimizing what we do experience as increased militarization. I would like to give an example of that. General Huddleston this morning referred to the cadets in many of the Inuit communities and said that this is an organized youth activity and, in some cases, the only organized youth activity. You also referred, I think, to the Rangers. In these communities there are no viable, self-reliant economies. I am suggesting that for the military to provide the only organized youth activity is, to say the least, not the most effective way to be approaching sovereignty issues in the North. So my specific question is: Why did you give no emphasis in your talk to the importance of developing viable economies and viable, self-determining groups in the North as part of our sovereignty objective? Would you comment on that, please?

Harriet Critchley
Yes, I would be happy to. First of all, I am sorry if you thought that I was justifying the increased militarization of the North. I had absolutely no such intention whatsoever of doing that, because as far as the Canadian part of the Arctic is concerned, I don't agree that there is increased militarization. I think it has decreased since the 1950s.

Second, I didn't talk about cadets. I did talk about Rangers. I have met some Rangers in various Arctic communities in Canada's North. I haven't met them all, of course, but my distinct impression is that, first of all, they are adults. Second, they are usually the one or two best hunters in the community, and it is regarded by them as an honour to join the Rangers; they are proud of it. They are not, I would suggest, underemployed at all. In fact they are probably among the more highly regarded members of the community because of their skills on the land.

You mentioned that I gave no emphasis to political development in the North. I'm sorry that you had that impression, because that was one of the four major factors I discussed at the outset of my remarks. I think northern political development and all that comes with it in terms of increased political awareness on the part of northerners—increased political activity on the part of northerners at the local level, at the national level, and at the international levels of politics, their increased participation in debates and in planning concerning economic development, concerning potential trade-offs, concerning a greater voice in the constitutional debate in Canada—are all very evident and, I think, extremely important. That's why I isolated that as one of the four factors I talked about. I think it is having a major impact. We see this over the past day and a half even here. It is having a major impact on how we—now I am speaking as a southern Canadian—see the North and how we think about the North. I think this is a long overdue and very welcome impact. I think also, for all of us as citizens in a democracy, that the very least we can do is to participate in these debates and make our views known and try to co-opt others into a more Canadian view. In that sense, I would say that this is an assertion of Canadian sovereignty—making known the values we hold and discussing them and reaching some consensus on them.

Phil Ralphs (Edmonton, Alberta)
Mr. Shallhorn, I am sure your talk was really informative, but I have a little question here. If the North is to be strong and free then we must have truth. Isn't it true that we would have a much more peaceful and environmentally sound world if the people who have lived in the North for thousands of years were in truth allowed their freedom? And the second part of that question is: Isn't it only then that we all could be strong and free?

Steve Shallhorn
I'm not entirely sure what was being asked. However, I think my general sense would probably be to agree. I was having trouble hearing the question.

Adrienne Clarkson
It is basically a question of the rights of the native peoples, their right to determination and their right to continue their way of life as they have always lived it and to play a role in the development of the future.

Shallhorn
That's essentially correct.

Ralphs
That is right, because they understand what the balance of the environment is all about, so if we left it in their hands they would know what to do. And they will not have any trouble getting resources because they also know how to do that. If they were in control of it politically and economically they would know how to use it wisely, whereas we don't.

Richard Murray (Edmonton, Alberta)
I would like to direct my question first to Mr. Shallhorn and then to Dr. Critchley, if she has some comment on it. It is not a very specific question. I was hoping you could perhaps inform me more about the involvement of special interest groups, such as corporate interests. Just how are they involved in the decision to develop cruise missiles or to purchase nuclear submarines? Could you give me some basic information on that and if possible direct me to some sources where I could find out more about it?

Shallhorn
I think right now in the United States—it was part of the underlying politics of the John Tower debacle,[1] I think—the question is what role industry, especially in the United States, is going to play in the determination of the level of defence spending that the United States maintains. I think that the indications are that industry is going to have less impact and less influence than it did during the previous eight years—the Reagan administration. I think it is clear that the U.S. military budget is about to shrink. I think the defence industry still continues to have a lot of energy, a lot of momentum, in the United States that will continue to propel it, but I am actually quite optimistic that we are going to see a slowing down of that momentum.

As to the influence of industry in Canada, I think Project Ploughshares is probably the organization that has done the most research on that, and there is a book published by their director, Ernie Regehr, called *Arms Canada* that would probably be the best source for you.[2]

Question Session

Jim Rodgers (Alberta Trappers Association)
Mr. Shallhorn, I have been representing the Trappers Association in Fort McMurray with regard to the oil sands development and other effluent impacts on the environment. We are really deeply involved in trying to defend the habitat and wildlife and the clear water and the waters that are running into the Arctic Ocean. We have been struggling somewhat in a vacuum. It would appear that our governments are content to allow industry to monitor itself, and industry's monitoring, although it might meet their requirements, could be seen as somewhat suspect. That leaves us in a position where governments cannot have any idea of what is flowing into the Arctic Ocean, for example, because they don't really have any control over the veracity of the information they are getting from industry. I'm sure that a lot of industry doesn't have the track record of the Tobacco Institute, but I can't imagine why we should trust other effluent producers more than those. We would like to have the assistance and cooperation of major environmental concern groups like Greenpeace, but we have been having a problem because of some generic misunderstanding about wildlife management and habitat preservation. Do you think there is any hope of our getting any assistance in the future?

Shallhorn
Yes, I think there is, and I think you will find that within Canada Greenpeace is quite interested in working with groups such as yours, including trapping organizations.

Rodgers
I would certainly welcome any initiatives. In fact, I'll make the first step.

Notes

1. The reference is to John Tower, the former Texas senator, who failed to obtain congressional confirmation of his appointment as President Bush's Secretary of Defense. His previous relations with defence contractors were presented as evidence against his suitability for the appointment. (Editor)
2. Ernie Regehr. *Arms Canada: The Deadly Business of Military Exports*. (Toronto: James Lorimar & Co., 1987).

CHAPTER 27
Summation

GURSTON DACKS AND FRANKLYN GRIFFITHS

Gurston Dacks

I am not in any sense a specialist on strategic studies or foreign policy. My interests are primarily domestic, internal to Canada and its North.

In thinking about the themes of this inquiry over the past two days, I have chosen as a metaphor another event that has been happening this week, the World Figure Skating Championships. It is an event which, I think, has attracted particular attention in Edmonton to the performances of our single skaters, but I am thinking more about the pairs. What is wonderful about the pairs is the way they diverge and go their independent ways, and then like magic they come together to produce the desired effect. What I take away from this inquiry is that this is precisely not what is happening in the North. What we see in the North, in several senses, is that there is a variety of processes under way, but these processes are not adequately choreographed. The failure of the choreography is something that I think causes us, or ought to cause us, significant concern. I see four pairs of processes that aren't being choreographed very well.

First, Mary Simon, John Merritt and Tom Berger talked about the process of environmental degradation and destruction, and they suggested that the process is proceeding faster than is our ability to control, monitor and improve the Arctic environment. Indeed, our efforts are accelerating, but they are not accelerating fast enough, and thus we face a very severe jeopardy.

Second, we have heard Willy Østreng and others say that there has been a process of militarization in the Arctic, which has not been matched by our efforts to bring that militarization under control or to reach a

Summation

Gurston Dacks

shared understanding of how we may secure our lives, our futures, with regard to that militarization.

Third, Gordon Robertson talked about northern political development. He noted the growth in the governments of the two territories and advocated enhancement in that growth, and I think that that is all to the good. I also sense, however, that there is another process, and that is the process of aboriginal self-government. It is not clear to me that the process of aboriginal self-government has been able to keep up with the growth of the territorial governments. My concern is that ten or fifteen years down the road, we may find the territorial governments so strongly empowered that there is rather little room for aboriginal self-government in the North. The concern that I then have is that the government of the Northwest Territories may dispose itself towards aboriginal people as unsympathetically as the government of Alberta does, and I do not think that is good enough. This is a possibility because the development of special vehicles for expressing aboriginal concerns is lagging, even though the government of the Northwest Territories relates to a population that is predominantly aboriginal—or at least, according to the 1986 statistics, equally aboriginal and nonaboriginal.

Finally, it seems to me that although there is an explosive process of growth in the significance of northern issues, yet we do not see in the policy of the government of Canada, despite some encouraging initiatives, the development of a coherent, integrated and explicit Arctic policy. We don't see, it seems to me, the linkage between the Canadian government's non-military policies—domestic policies and policies of international cooperation and exchange—and the military policies. And it seems to me that if we are going to make the best of the opportunities that have been suggested to us in the past two days, we need this kind of coherence.

I think that if we want to bring these four pairs of processes into coherence—if we want to get them to the conclusions that we would like them to accomplish—we have to think of the figure skating metaphor. We

Gurston Dacks is a professor in the Department of Political Science at the University of Alberta and an adjunct research professor at the Boreal Institute for Northern Studies there. His current research concerns the devolution of administrative responsibilities from federal to territorial governments. He has also studied the prospect of the division of the Northwest Territories and alternative forms of government for the western N.W.T., should division occur. He holds a bachelor's degree from the University of Toronto and a doctorate from Princeton University. He was appointed to the faculty of the University of Alberta in 1971 and is author of A Choice of Futures *(Toronto: Methuen, 1981), a textbook on the politics of northern Canada.*

Summation

have to recognize that the genius of the skaters is not so much what they do on the ice but rather the work of the choreographer. That is what Canada does not now have, it seems to me—a government choreographer.

At present, we have more than a dozen government departments actively pursuing northern initiatives, each primarily concerned with its own mandate, often working at cross purposes. General Greenaway mentioned the conflict between the departments of Transport and National Defence. With regard to land claims, the departments of Environment and of Fisheries and Oceans have been at odds with the Department of Indian Affairs and Northern Development concerning the need for aboriginal people to share in the management of wildlife. That has delayed the settlement of land claims, which is crucial if the integration of the aboriginal agenda and the broader northern agenda is to come about. The Department of Energy, Mines and Resources is in conflict with the Department of Indian Affairs and Northern Development regarding the transfer of powers to the territorial governments.

What I suggest Canada needs is a choreographer. It needs a Minister of Arctic Policy. That is what the government of Canada needs—not a Minister of Indian Affairs and Northern Development, because his or her mandate is relatively narrow in terms of function and because the existing department is in the process of being downsized, if not really phased out, with regard to its northern responsibilities. So we need a minister, not of an enormous department, but a minister who has the ability to make the other departments pay attention, to coordinate their policies with regard to a focus on the North. That means that he or she needs support in both the Privy Council Office and the Prime Minister's Office.

The only way we can pursue a northern policy is if the various departments are compelled to follow that policy rather than to go in their own directions. Beyond that, I think we need to integrate, or to involve, the public in a meaningful fashion. It is very interesting to note that Canada has a consultative committee on strategic affairs, but does not have a consultative committee on Arctic policy. Clearly, if the government of Canada, in the last five years in particular, has been working on means of involving the public in consultative committees—and that is what they did with the free trade arrangement, whether you like it or not (they had seventeen consultative committees or some such number)—if that is the way they work, why isn't there a consultative committee on Arctic issues? And if we have a strategic consultative committee, why is it the case that there are, to my understanding, no northerners on that consultative committee?

The government of Canada needs to be thinking coherent, comprehensive Arctic policy. My understanding is that in the Department of External Affairs there is one person with defined responsibilities for Arctic affairs, and I submit that one person, however excellent, is not an adequate reflection of the priority that we ought to be attaching to Arctic concerns. My sense of the matter is that we need to be doing all of these things.

Saying that "we" need to be doing all of these things is sometimes a rhetorical device because the question is: Who is the "we" about whom we are talking? That raises the question: What is the function of an inquiry such as the one in which we are engaged? It seems to me that the function of the inquiry is to educate people who attend and to cause them to reflect, but also ultimately to cause some change in government behaviour, if that is what we are looking for.

My sense of the past two days is that I see an alienation from the government of Canada expressed in the questions, in the responses of the audience, and so on. My concern and my hope would be that what you take away from this inquiry is not alienation, but rather a sense of the need for something that might be called "constructive engagement." I think what we need to do if we are not satisfied with the answers of our government is not merely to disdain them, but rather to work through groups such as the groups represented in the foyer outside this room.[1] We need to engage the government of Canada to develop the expertise to speak to such groups at the level at which the government makes its policies and to respond to interest groups generally.

We need to compel the government to rethink its approaches to the North, in the kind of sophisticated way that government policies are actually made, to see that what is necessary is a northern focus. We need to compel them, when they have gained that focus, to see, for example, that land claims, far from being an obstacle to the Department of the Environment and the Department of Fisheries and Oceans, should be a major instrument by which we, the people of Canada, demonstrate our historic occupation of the North. We understand that environmental protection requires us to focus in the North and that we need to view these questions of regional or global demilitarization from a different perspective than is currently the case. I urge you to think about that: not just to be sceptical of our government's officials, but rather to plan how to engage them constructively so that they will both increase their focus on the North and rethink their priorities with regard to it.

Franklyn Griffiths

I'm flying by the seat of the pants in making these comments. I come away from this inquiry with many interesting impressions, a great number of ideas. There is much to handle by way of summation.

I study the Arctic but am not an Arctic expert. I write about the area and travel there occasionally. I see myself—I think for the most part we

Summation

Franklyn Griffiths

think of ourselves—as southern Canadians who are interested in the Arctic, who are concerned with its troubles and its problems. I have learned a great deal and I hope you have as well.

I would say, by way of additional introduction, that I am a Canadian nationalist. I am also a Canadian monarchist, which goes perhaps even further. I am a monarchist flying by the seat of his pants.

Right off the top, I feel awkward because there is no northern native person taking part in this wrap-up. Here we have two southern non-natives talking about the North. We should learn this is not the way to do things. Next time around it should be different.

Our policies in the North will be right if our attitude is right. The important thing is to get the right attitude, the right thinking. I start to see some of the right thinking at this meeting. I see it elsewhere in Canada. I see signs of attitude change that are rather promising. They are not what they might be and should be, but I would flag the way in which we have discussed the rights of northern native peoples, the need to empower them. This is a very important concern, and incredibly I hear it voiced elsewhere in Canada. This is all to the good. There is also the recognition of the environment and other critical aspects of international, as well as national, Arctic affairs, which we have discussed here.

One particular aspect I do want to stress is one in which I come away from this meeting with my own thinking changed a bit. This is the question of sovereignty. Canadian attitudes about sovereignty are starting to alter. Indeed, I would say that it is getting near time to put a ban on the use of the word sovereignty. That may sound a little radical to you, but let me try to explain.

As a people—I refer to southern Canadians—we are still living with some rather ancient and antique ideas and feelings about the Arctic deep in our minds and hearts. These attitudes probably come from the nineteenth century and ultimately from the history of British exploration before Confederation. They constitute a vision of the Arctic sublime, in the

Franklyn Griffiths is a professor of political science at the University of Toronto, where he has taught and written since 1966. At various times, he has been director and acting director of the university's Centre for Russian and East European Studies. In 1986–87 he was senior policy advisor in the office of the Secretary of State for External Affairs. The following year he was visiting professor at Stanford University. In addition to the Arctic, his research interests include Soviet affairs and arms control. Mr. Griffiths' publications include, as editor, Politics of the Northwest Passage *(Kingston and Montreal: McGill-Queens University Press, 1987) and, as co-editor with John Polanyi,* The Dangers of Nuclear War *(Toronto: University of Toronto Press, 1979).*

original sense of that word. That is, the Arctic is able to raise us above our ordinary selves, to give us a sense of exalted being, to make us marvel at the beauty it presents. But at the same time it is terrifying. It is awesome in its mindless destructiveness. It inspires fear. It is at once appealing and appalling. If we look in our own minds today, there is still a sense that the Arctic is sublime. I would say this of myself and my own attitude: the Arctic is magnificent and yet I am taken aback by it. I treasure it, I feel very strongly about it, I feel that it is a vital part of what makes me a Canadian, of what makes Canada different from other countries and of what lies ahead in Canada's future. All this and more is inherent in the notion of the Arctic.

And yet, at the same time, there is the element of not going up there. Why haven't I spent more time in the Arctic? Why haven't you all? Why haven't we? Why do we go south for vacations? There is something in our mentality that holds us back and drives us away. The climate is an obvious consideration. But there is a deeper aspect to it—a Canadian dualism we continue to live with that makes our minds divided where the Arctic is concerned. This dualism has worked against us for decades. We have never really examined these thoughts and feelings of ours. It is high time we did.

Our traditional attitude is evidenced by a fitfulness in the way we southern Canadians approach the Arctic. On the one hand, when our Arctic sovereignty is threatened in any way, especially by the United States, we rise up in high dudgeon. We become very, very concerned. However, as soon as the threat is gone, we go to sleep about what needs to be done in the region; we do not demand things of our government. And the government on the whole tends to steer its listless northern course.

On the matter of sovereignty, though, I do see change and I hear some words of change at this inquiry. We are voicing a vision and a way of looking at the Arctic that is beginning to alter. We are not thinking so much as before about legal lines, with ourselves behind the lines trying to keep people out who are trying to break through. We are instead recognizing that we are a distinct, and perhaps unique, people in an interdependent world. We have a separate Arctic destiny that is part of our larger destiny; but really the world is interdependent. We must share, we must be open, and the Arctic, whether we like it or not, is open. It is open to pollutants. It is open to climate change—climate change that, if some predictions come to pass, is going to destroy or very, very seriously threaten the cultures of northern native peoples, not only in Canada but throughout the circumpolar North.

We are increasingly aware as a people that we need to look more outwards, that we cannot, as it were, stay back behind the lines defending sovereignty. I regard this as a positive trend. If I were to put words to an inherent new vision that may be starting to come to Canadians as they consider the Arctic, there is one particular word that I would stress. It is civility. It seems to me that we are moving towards a new recognition, towards showing due respect in ways we have not shown it before. We are moving towards showing respect, first of all, to our physical environment in the Arctic, where we have all too often sought to overrun it, and where we have not accorded it the respect that it is due. Something is changing here. We are moving towards showing due regard and honour to our native peoples, whom we have not treated with respect. Civility applies here as well.

Furthermore, turning to international relations, we are in need of greater civility among adversaries and towards our Soviet adversaries. This greater civility is starting to show in the West during a time of renewed détente, and we see the same from the Soviet Union. Less and less are we treating one another with incivility, that is, making accusations and brandishing nuclear and conventional weapons actively. More and more we are treating one another with civility and respect. There is an important dimension of change here in the Arctic that we should take note of, and I see it at this inquiry.

More broadly, I see two agendas in the development of circumpolar affairs. There is the old agenda of international security affairs that is still important and will be with us for some time to come. It emphasizes questions of hardware, strategy, numbers, who's got what warheads, nukes and the rest. There is, however, a new agenda that is overtaking the old, and the future lies with the new agenda. The word very often used in this inquiry to describe the new agenda is inaccurate; the word employed is "non-military," as opposed to "military." We should not be talking about this new agenda as a negative, as non-military. The right word for it is "civil." And again I come back to civility. We are talking about international environmental *cooperation*, cooperation in human rights and the preservation of native cultures, cooperation in scientific affairs and the rest. This is civil behaviour. It is not simply non-military. There is more to it than that. The future lies with this kind of activity.

There was an interesting passage in Willy Østreng's discussion yesterday about whether or not there was still a united or coherent security concept in the Arctic as we look ahead. He suggested, and I think rightly so, that on the whole an integrated Cold War security concept has been disassembled and we have something rather more diffuse now. I ask though, and I think the Soviets are also inclined to ask, Is there still an idea, a concept, that can help guide us in our international Arctic affairs and to some degree in our domestic affairs as well? The two are in many ways inseparable. "Security" is not the overarching word that describes everything we aim for at home and abroad. I say "civility" is. We should not be

Summation

talking as though security were something you can have for yourself. We should be talking about desirable relationships between Arctic countries, between southerners and northern natives, and between all of us and our environment. The word that describes these relationships is not really "security." The word is "civility." It is also part of Canadian political culture, our way of looking at things. We should act on it.

Some other points: one of them has to do with the empowerment of our native peoples and their capacity to participate in domestic life as well as in international Arctic affairs. The same applies to the Alaskans, to Greenlanders and, to a degree, also to the Soviets. The native peoples are not going to be given a greater role in political life at home and abroad simply and primarily because we recognize that this is just, this is right, this is civil. It is going to be more complicated. Natives are going to gain a greater role because people in southern decision-making centres recognize increasingly that the Arctic has an ever greater strategic military value and that it therefore needs political stability. If the native peoples choose, as the Inuit Circumpolar Conference does, to call for an Arctic zone of peace, they will defy the established military priorities of southern governments. Already they are posing a challenge that will have to be answered. And the answer is going to have to be: Pay more attention to, listen to, and not merely listen to, but actually take the advice of, northern native peoples.

This, by the way, applies to the Soviet Union. Our Soviet guests have not told us this, but it is my understanding that in the Soviet Union right now, in the Soviet Arctic, there is just about a total ban on new megaprojects, on big new industrial developments. The Soviet northern native peoples are able to say no. This is a plus. The problem is not one that affects only the West. It is a circumpolar problem and a part again of the larger interrelationship.

A few final thoughts about specific things that might be done. First, I believe there are a few requests we should be making of the Soviet Union. In a way, it seemed to me that they got off a bit easy in this inquiry. The Soviet Union retains in readiness a nuclear testing site in Novaya Zemlya in the Arctic. This is a Soviet island high up in the Arctic Ocean. They have the capacity there, if they choose, to start nuclear testing in the Arctic. It seems to me this is absolutely wrong. We should be saying to them, and others should: How can you have a credible policy of peace for the Arctic and seek to turn it into a zone of peace, when you are ready to start testing nuclear weapons there at the drop of a hat, when you feel like it? They should be invited to dismantle that site as a token of good will. If they are going to do their testing, do it outside the Arctic. Let's get the Arctic nuclear-free in that respect, and the Soviets should start first on that one.

There is another thing that troubles me about Soviet actions, and this is a particularly Canadian concern. The Soviets fly long-range strategic bombers at Canadian air defences, especially in the Beaufort Sea. This to me is a threatening gesture. They do not need to do it. I don't like it happening, and I don't like what it has allowed the Department of National Defence in Canada to do, that is, to infer that the Soviets are threatening, that you cannot cooperate with them, that you need military strength, that you need strategic nuclear submarines. We should get to the root of the problem here and ask the Soviets to stop flying their strategic bombers against Canadian air defences. That is something they could surely do to be more civil.

There is one other thing I would mention. Ambassador Roche cited Canada's long and distinguished record in the field of international peace-keeping. We should build on that record. Canada is in a good position to show leadership in proposing the establishment of an international environmental emergency force. This would be a force consisting of national units on a stand-by basis, air-transportable, very largely military but with paramilitary elements as well. It would be able to deal with oil spills, let's say, in Antarctica; to get down there right away when a Chilean tanker goes on the rocks; able to deal with Arctic environmental disasters as well; and able to respond to, rather than simply study and think about ways of adapting to, the environmental degradation that surrounds us. Canada has a role to play here. The money for this kind of force should come out of military budgets for capital improvement. It should be a straight transfer from the military to the civil. Again, I am talking about civility. I leave it at that.

Notes

1. A large number of the organizations and institutions sponsoring or endorsing the inquiry, including public interest groups and government departments, were dispensing information from tables in the foyer during breaks in the meeting. (Editor)

CHAPTER 28
Closing Remarks

BRIAN SPROULE

We have received a communication from Chief Daniel Ashini of Labrador, which, I think, represents some of the very real concern and passion that the people of the North have about their country. As a postscript to the inquiry, and in line with some of the remarks that Dr. Griffiths has just made regarding participation by northern native people in the inquiry, I propose to read it. It is as follows:

> My warm greetings to The True North Strong and Free Inquiry in Edmonton. I wish you great success for your conference.
>
> We have an issue in Labrador and Quebec which we would like the participants of your conference to consider for their support.
>
> For thousands of years the Innu have lived quietly on the Quebec/Labrador peninsula, in their homeland which they call Nitassinan. They fish, trap and hunt caribou, the mainstay of their physical and spiritual lives.
>
> There are 10,000 Innu, as the Naskapi-Montagnais people call themselves, in Nitassinan. Each spring and fall, Innu from communities like La Romaine, Mingan, Natasquan, Ste. Augustine, Davis Inlet and Sheshatshit travel far into the bush to hunt and fish. Here the knowledge of hunting, trapping, fishing, hide preparation, cooking, snowshoe making, and traditional beliefs and rituals are passed on to the children. Without this seasonal migration, lasting sometimes eight months, the Innu culture would cease to exist.
>
> Our traditional way of life is threatened by the military development taking place over our land.

Our ancestors cared for Nitassinan so well that a 1985 issue of *The British Air Force News* said: "The real beauty of this place is that, quite simply, the land is just as God left it."

The Government of Canada is destroying the beauty of our land. We have never signed a treaty with Canada, so the land is ours. Yet the Canadian government has given West Germany, the Netherlands and Great Britain permission to fly low-level military planes over our land and to use it for bombing practices. They allow the military flights thirty metres above the ground. This is completely forbidden and impossible in Europe, where the low-level flights are only allowed 150 metres above the ground.

Daily our people and the animals we hunt are hurt and frightened by screaming jets which fly as low as thirty metres over our heads. Time after time, since 1980, we protested this illegal activity to the Government of Canada, but we were completely ignored. We feel much like we think you would feel if the Government just took your house away and shoved you and your children into the street.

Listed below are more of our concerns regarding the military development.

- Innu children and elders have had very traumatic experiences with overflights. When in our hunting camps, children, when overflown, have been seen to run into the woods in fright, and at times have been lost for two or three hours. Therefore many children have refused to go with their parents to their hunting camps.
- The illegal seizure of our lands for a practice bombing range—where we have found evidence not only of sophisticated laser-guided bombs, but also live ammunition—is in violation of the military guidelines for the range.
- Continued and escalating low-level flight exercises over Innu lands and camps. In 1988, there were 7,500 flights from Goose Bay, which is an increase of thirty-three percent over 1986 flights.
- The growing occurrence of sonic booms over Innu people in the country, four of which have been reported in the past.
- The pollution of our homeland with PCBs, asbestos dumping, and the dangers of hydrazene from Dutch F-16s.
- The proposed NATO centre, which, over the next ten years, will turn our whole homeland into a gigantic military playground.
- Since these exercises have commenced, a number of crashes of these aircraft has occurred. It is our grave concern that the next crash may very possibly take place on one of our camps.

- Since these aircraft are capable of carrying nuclear weapons, we feel this is a great risk to world peace.

European governments have recognized that low-level flights affect the health and environment of their people. Therefore they export their exercises to our land and our people because they expect much less resistance. Why is the health and the environment of the Innu considered less valued than that of Europeans, or not valued at all?

Our voice was not being heard through the campaign of letter writing, press releases, meetings and low-key demonstrations. Therefore in September of last year we decided to step up the fight. We set up camp on the Minipi Lake bombing range south of Goose Bay, halting all bombing practice for two months. Eight times through the fall we breached base security by walking through the airport gate, over the fence, past F-4 Phantoms, F-16s and Tornados, and parked ourselves on the tarmac, led by our elders in prayers and hymns. Training flights were disrupted and 115 charges were made.

This spring we will do the same and continue our struggle.

We hope the True North Strong and Free Inquiry will recognize the issue and support us in our struggle.

I think that represents a plea and some of the feelings that are obvious out there in the people of the North.[1]

I stand here as a sort of a historical artifact, the tattered remnant of a bygone inquiry, The True North Strong and Free? Inquiry of 1986. My old bones have been absolutely thrilled by what I have experienced here in the past two days. I did predict and did anticipate they would be, and my expectations have indeed been met.

This wonderful event has been part of an ongoing process. It all started in January of 1986 when Laurie McBride and a small group of citizens concerned about nuclear submarine activities in the inland waters of the Pacific coast at the Nanoose Bay Naval Testing Range organized a public inquiry in Nanaimo. The event evoked quite an astonishing response. Mel Hurtig of this city attended, and he returned home fired with enthusiasm that he used to energize a number of individuals and groups to mount an inquiry in Edmonton focusing on broader issues of Canada's defence policy and nuclear arms. It was co-sponsored by the Northern Alberta Chapter of the Council of Canadians and the Edmonton Chapter of Canadian Physicians for the Prevention of Nuclear War. On November 8 and 9, 1986, fifty-five hundred bone-chilled souls battled their way through the year's first blizzard to a university athletic stadium to be educated, challenged, stimulated and motivated, just as we have been here over the past two days. The same exceedingly effective format has been used for all three inquiries, emphasizing in each instance the articulate delivery of balanced information coupled with challenges, first from a knowledgeable panel and then from the public.

Canada's geography makes concern over the utilization and management of our fragile northern shores an imperative to us all. The necessity for circumpolar cooperation in addressing urgent environmental concerns against the background of military posturing reflects, in microcosm, problems that encompass the entire planet. As in the threatened North, which has been so thoughtfully considered at this inquiry, activities for reasons of profit, short-term and long-term employment, or perceived or real defence needs may all result in actions that will irretrievably damage the home in which we live, our precariously balanced environment. It is entirely appropriate that these issues should be publicly considered and examined by a broad spectrum of the populace. They are far too important to be left only to politicians or only to the military or only to the scientists—or only to the environmentalists for that matter.

For my part, the last two days have been a great experience. On behalf of those who have attended I would like to thank and to commend the organizing committee, who are listed in your program, and most particularly the program committee, who brought together the wonderful, knowledgeable and articulate speakers—the careful and cautious Canadians, the peace-loving Scandinavians, and the daring and imaginative Soviets—who have elevated the spirits and the level of understanding of us all.

This inquiry, however, is not over. The most important aspect starts as we leave this building. It is now for each of us to distill a personal message from what we have heard and then to act on it: to volunteer to meaningful organizations; to transmit our concerns to local and national political representatives; to contribute time, money and energy to those people or groups that can make a difference, that can transform concern into action.

We will see you all at the next True North Strong and Free inquiry.

Notes

1. The Innu (Montagnais, Naskapi) of Labrador and northeastern Quebec, reported to number about nine thousand, have an unresolved land claim over an area of some 259,000 square kilometres (the area of Labrador itself is some 292,000 square kilometres). The claim area includes a Canadian Forces installation at Goose Bay, used as the base for low-level military flight practice. As a result of sit-ins, 219 Innu have been arrested: of these, 4, including the local chief, Daniel Ashini, were tried on and acquitted of charges of public mischief in April 1989. The provincial court judge viewed the defendants as acting under the honest belief that they possessed hereditary rights to the land on which the base is situated. (Editor)